Textbook of Organic Farming

THE AUTHORS

Mr. Amit Kumar Singh obtained his M.Sc. (Ag) Agronomy from Acharaya Narendra Deva University of Agriculture and Technology, Kumarganj, Ayodhya (U.P.). He has qualified ARS (ICAR-NET) conducted by Agricultural Scientists Recruitment Board, New Delhi. Currently working as Block Technical Manager (BTM), ATMA, Paliganj, Patna. He worked as Senior Research Fellow (SRF) at ICAR-Research Complex for Eastern Region, Patna in research project sanctioned by Bill Gates Foundation, United States. He has vast experience in popularization of Agricultural farming techniques in the farmer field. He has published two books and number of research papers in national and international journals of repute and also organized/ participated various training programmes, seminars, symposia and workshops.

Dr. Vinod Singh is presently working as Associate Professor in the Department of Genetics and Plant Breeding, College of Agriculture, Acharya Narendra Deva University of Agriculture and Technology, Kumarganj, Ayodhya (UP). Dr. Singh has more than 32 years of experience in teaching, research and extension. He has published various research papers in journals of repute and guided several M.Sc. (Ag) and Ph.D. students. He has written more than 13 books and successfully guided several M.Sc. and Ph.D. students. Dr. Singh has been honored with Fellowship viz. Fellow International Society for Environmental Protection (FISEP) and Fellow Living World Society (FLWS) and he has delivered several talks on Radio and Doordarshan on different aspects of Agriculture.

Textbook of Organic Farming

– Authors –

Amit Kumar Singh

Vinod Singh

2024

Daya Publishing House®

A Division of

Astral International Pvt. Ltd.

New Delhi – 110 002

ISBN: 9789359191874

Publisher's Note:

Every possible effort has been made to ensure that the information contained in this book is accurate at the time of going to press, and the publisher and author cannot accept responsibility for any errors or omissions, however caused. No responsibility for loss or damage occasioned to any person acting, or refraining from action, as a result of the material in this publication can be accepted by the editor, the publisher or the author. The Publisher is not associated with any product or vendor mentioned in the book. The contents of this work are intended to further general scientific research, understanding and discussion only. Readers should consult with a specialist where appropriate.

Every effort has been made to trace the owners of copyright material used in this book, if any. The author and the publisher will be grateful for any omission brought to their notice for acknowledgement in the future editions of the book.

Published by : **Daya Publishing House®**
A Division of
Astral International Pvt. Ltd.
– ISO 9001:2015 Certified Company –
4736/23, Ansari Road, Darya Ganj,
New Delhi-110 002
Ph. 011-43549197, 23278134
E-mail: info@astralint.com
Website: www.astralint.com

Preface

Organic farming in simple terms is a kind of agricultural practice where harmful chemical fertilizers, pesticides and insecticides are not used. It is primarily based on developing and cultivating land using only natural input such as soil, water, and air. The purpose of organic farming is to develop healthy and nutritious crops as well as save our environment from degradation and pollution. Organic farming helps countries achieve sustainable use of natural resources. Non-renewable natural resources such as water and coal are increasingly used by the agriculture industry and hence inculcating organic farming will help save our natural resources. The trend of consumption of organic food is increasing every day in the world market. Once the consumers get to know the health benefits of organic food, then the organic agriculture market will start to boom. Organic produce is in high demand today because of the health-conscious consumer. Consumers are aware that toxic chemicals and genetically modified organisms as chemical fertilizers are used in growing crops which can have negative health effects on them and as a result of which people around the world are increasingly consuming products of organic farming.

The present day for self-sufficiency in food grain production may not last longer unless we develop a sustainable agricultural system which maintains and/or improves soil fertility and productivity with greater acceptance of biological principles so as to assure adequate/more food production in future. Besides plants are more prone to pest and diseases in intensive agriculture, use of chemicals can have residues on the produce, in the soil and in ground water. With more purchased inputs, the cost of production is also mounting up. Pesticides used in paddy, cotton and vegetables which occupy less than 30 percent of total area account for more than 80 per cent of the chemicals used. Organic farming practices that reduce the pressure on land, water and biodiversity without adverse effects on agricultural production and nutritive value of food comprise judicious use of organic manure, *viz.* Farm yard manure, compost, crop residues, Vermicompost *etc.* integrated is an efficient nutrient management practices, cropping systems, conjunctive use of rain, tank and underground water, integrated pest management and conservation of genetic resources. Among them, soil fertility is given top attention due to its dynamic action with various physical, chemical and biological properties.

This book has been designed to cover undergraduate syllabus of *Organic Farming* as per ICAR 5th Deans Committee report. It covers the entire syllabus in one compact volume of 11 chapters viz. introduction of organic farming, principles of organic farming, concept of organic farming, certification, labelling and accreditation process, soil improvement and amendments, weed management in organic farming, organic manure, vermicompost, integrated disease and pest management, marketing and exports of organic products and biopesticides. Related terms have been provided at the end for the ready reference of readers. The present book has been recommended as a core subject in the curriculum of B.Sc. (Agriculture) course of the agricultural universities. This book would be immensely useful to students, teachers, researchers, progressive farmers and extension workers working in the field of crop science, agronomy and organic farming.

We express indebtedness to authors of books, manuals, periodicals, magazines, newsletters, journals and websites *etc.* that helped in the preparation of this book. It has not been possible to obtain permission for reproduction of some of the materials for which authors and publishers offer their sincere apologies. Healthy criticism and suggestions for further improvement of the book are solicited.

Amit Kumar Singh

Vinod Singh

Contents

Chapter 1

Introduction

1.1 Introduction: An Overview

Sustainable development has caught the imagination and action all over the world for more than a decade. Sustainable agriculture is necessary to attain the goal of sustainable development. According to the Food and Agriculture Organization (FAO), sustainable agriculture "is the successful management of resources for agriculture to satisfy changing human needs while maintaining or enhancing the quality of environment and conserving natural resources". All definitions of sustainable agriculture lay great emphasis on maintaining an agriculture growth rate, which can meet the demand for food of all living things without draining the basic resources.

Organic farming is one of the several approaches found to meet the objectives of sustainable agriculture. Many techniques used in organic farming like inter-cropping, mulching and integration of crops and livestock are not alien to various agriculture systems including the traditional agriculture practiced in old countries like India. However, organic farming is based on various laws and certification programmes, which prohibit the use of almost all synthetic inputs, and health of the soil is recognised as the central theme of the method.

Adverse effects of modern agricultural practices not only on the farm but also on the health of all living things and thus on the environment have been well documented all over the world. Application of technology, particularly in terms of the use of chemical fertilizers and pesticides all around us has persuaded people to think aloud. Their negative effects on the environment are manifested through soil erosion, water shortages, salination, soil contamination, genetic erosion, *etc.*

Organic farming is one of the widely used methods, which is thought of as the best alternative to avoid the ill effects of chemical farming. There are several definitions of organic farming and the one given by the US Department of Agriculture (USDA) is considered the most coherent and stringent. It is defined as 'a system that is designed and maintained to produce agricultural products by the use of methods and substances that maintain the integrity of organic agricultural products until they reach the consumer. This is accomplished by using substances, to fulfil any specific fluctuation within the system so as to maintain long term soil biological activity, ensure effective peak management, recycle wastes to return nutrients to the land, provide attentive care for farm animals and handle the agricultural products without the use of extraneous synthetic additives or processing in accordance with the act and the regulations in this part'. The origin of organic farming goes back, in its recent history, to 1940s. During this period, the path breaking literature on the subject published by J.I. Rodale in the United States, Lady Balfour in England and Sir Albert Howard in India contributed to the cause of organic farming.

The farming being practiced for the last three decades in India has increasingly been found non-sustainable. The system is oriented towards high production without much concern for ecology and the very existence of man himself.

1.2 Definition

The USDA defines organic agriculture as "a production system that is managed to respond to site-specific conditions by integrating cultural, biological, and mechanical practices that foster cycling of resources, promote ecological balance, and conserve biodiversity."

In another definition FAO suggested that "Organic agriculture is a unique production management system which promotes and enhances agro-ecosystem health, including biodiversity, biological cycles and soil biological activity, and this is accomplished by using onfarm agronomic, biological and mechanical methods in exclusion of all synthetic off-farm inputs".

Organic farming works in harmony with nature rather than against it. This involves using techniques to achieve good crop yields without harming the natural environment or the people who live and work in it. An organic farmer produces vegetables, fruit, cereal crops, or livestock without the use of chemical fertilizers, pesticides, or herbicides. In another way organic farming is kind of agricultural that provide the consumers, with fresh, tasty and reliable food while regarding natural life cycle systems. In addition to health benefits of organic products for consumers, there are vital environmental benefits for the earth. An organic farming keeps biodiversity and reduce environmental pollutions such air, water and soil. Organic agriculture has grown out of the conscious efforts by inspired people to create the best possible relationship between the earth and men.

More specifically, organic farming entails:

- ✰ Use of cover crops, green manures, animal manures and crop rotations to fertilize the soil, maximize biological activity and maintain long-term soil health.
- ✰ Use of biological control, crop rotations and other techniques to manage weeds, insects and diseases.
- ✰ An emphasis on biodiversity of the agricultural system and the surrounding environment.
- ✰ Using rotational grazing and mixed forage pastures for livestock operations and alternative health care for animal wellbeing.
- ✰ Reduction of external and off-farm inputs and elimination of synthetic pesticides and fertilizers and other materials, such as hormones and antibiotics.
- ✰ A focus on renewable resources, soil and water conservation, and management practices that restore, maintain and enhance ecological balance.

1.3 Milestone in Organic Farming

- ✰ **Sir Albert Howard (1900-1947)**: Father of modern Organic Agriculture, developed organic composting (mycorrhizal fungi) at Pusa, Samastipur, India and Published document 'And Agriculture Testament'.
- ✰ **Rudolph Steiner (1992):** A German spiritual Philosophical built biodynamic farm in Germany
- ✰ **J. I Rodel (1950):** Popularized the term sustainable agriculture and method of organic growing in USA
- ✰ **IFOAM:** Establishment of International Federation of Organic Agriculture Movement in 1972 in France.
- ✰ **One Straw Revolution:** Release of the book by MASANOBU Fukoka(1975), an eminent microbiologist of JapanEU regulation Codex guidance on organic standard 1999.

Table 1.1. Potential Benefits of Organic Agriculture in different Systems

Parameter	*Potential Benefits*
Agriculture	Increased diversity, long term soil fertility, high food quality reduced pest/disease, self-reliant production quality stable production
Environment	Reduced pollution, reduced dependence on non-resources, negligible soil erosion, wildlife protection, protection, resilient agro-ecosystem compatibility of production with environment
Social conditions	Improved health, better education, stronger community reduced rural migration, gender equality, increase employment, good quality work

Parameter	*Potential Benefits*
Economic conditions	Stronger local economy, self-reliant economy, income security, increased returns, reduced cash investment, low risk *etc*
Organizational/ Institutional	Cohesiveness stability, democratic organizations, enhanced capacity

1.4 Important Goals of Organic Farming

The important goals of organic farming are:

- ☆ A sufficiently high level of productivity
- ☆ Compatibility of cultivation with the natural cycles of the production system as a whole
- ☆ Maintaining and increasing the long-term fertility and biological activity of the soil
- ☆ Maintaining and increasing natural diversity and agro-biodiversity
- ☆ Maximum possible use of renewable resources
- ☆ Creation of a harmonic balance between crops and animal husbandry
- ☆ Creation of conditions in which animals are kept that correspond to their natural behaviour
- ☆ Protection of, and learning from, indigenous knowledge and traditional management systems

1.5 Relevance of Organic Farming

The relevance and need for an eco-friendly alternative farming system arose from the ill effects of the chemical farming practices adopted worldwide during the second half of the last century. The methods of farming evolved and adopted by our forefathers for centuries were less injurious to the environment. People began to think of various alternative farming systems based on the protection of environment which in turn would increase the welfare of the humankind by various ways like clean and healthy foods, an ecology which is conducive to the survival of all the living and non-living things, low use of the non-renewable energy sources, *etc.* Many systems of farming came out of the efforts of many experts and laymen. However, organic farming is considered to be the best among all of them because of its scientific approach and wider acceptance all over the world.

Organic Agriculture Statistics at a Glance in 2020-21

Area

Cultivated Area (Organic + In-conversion)	2657889.33 ha
Wild Harvest Collection Area	1681295.61 ha
Total Area(Cultivated + Wild Harvest)	**4339184.93 ha**

Source: https://apeda.gov.in/apedawebsite/organic/data.htm#Summary_Statistics_2021

Production

Farm Production	3468991.98 MT
Wild Harvest Production	27808.36 MT
Total Production	**3496800.34 MT**

Source: https://apeda.gov.in/apedawebsite/organic/data.htm#Summary_Statistics_2021

Operators

Individual farm producers	*3495*
ICS Groups	*4781*
Total Processors	*1703*
Total Trader	*745*
Wild Operators	*71 (Total collectors 8724)*
Total Operators	*10795 No.*
Total Farmers	1599010

Source: https://apeda.gov.in/apedawebsite/organic/data.htm#Summary_Statistics_2021

Organic Exports

Total exports quantity	888179.68 MT
Total Export Value (INR)	707849.52 Lakhs
Total Export Value (US$)	1040.95 million USD

Source: https://apeda.gov.in/apedawebsite/organic/data.htm#Summary_Statistics_2021

State-wise Cultivated Farm Area for the Year 2020-21

Sl.No.	*State Name*	*Organic Area (in ha)*	*Conversion Area (in ha)*	*Total Area (in ha)*
1	Madhya Pradesh	540993.98	479024.00	1020017.98
2	Maharashtra	219659.41	152063.21	371722.62
3	Rajasthan	177599.56	121086.74	298686.29
4	Gujarat	72318.02	75548.39	147866.41
5	Karnataka	61115.97	33934.11	95050.08
6	Odisha	78148.00	14546.81	92694.81
7	Sikkim	74647.31	1082.34	75729.66
8	Uttarakhand	31556.80	43269.61	74826.40
9	Uttar Pradesh	53194.79	14247.83	67442.61
10	Jharkhand	0.00	53261.70	53261.70
11	Kerala	25656.60	19413.78	45070.38

Sl.No.	State Name	Organic Area (in ha)	Conversion Area (in ha)	Total Area (in ha)
12	Meghalaya	34816.30	3560.09	38376.39
13	Andhra Pradesh	22343.07	14458.29	36801.36
14	Tamil Nadu	14086.32	17542.74	31629.06
15	Jammu and Kashmir	19028.70	11591.12	30619.82
16	Bihar	4.90	29897.64	29902.54
17	Chhattisgarh	14744.46	8465.06	23209.52
18	Assam	6719.27	11751.57	18470.84
19	Nagaland	7384.96	7405.42	14790.38
20	Arunachal Pradesh	265.37	12848.74	13114.12
21	Mizoram	40.45	12998.44	13038.89
22	Manipur	4419.25	8305.67	12724.92
23	Goa	9243.79	3388.53	12632.32
24	Himachal Pradesh	9108.87	2745.12	11854.00
25	Telangana	4723.74	2141.82	6865.56
26	Tripura	203.56	6317.75	6521.31
27	West Bengal	5462.79	839.82	6302.61
28	Haryana	3345.85	1557.22	4903.06
29	Punjab	879.87	1141.63	2021.50
30	Lakshadweep	895.51	0.00	895.51
31	LADAKH	0.00	817.85	817.85
32	Pondicherry	2.84	20.81	23.65
33	New Delhi	0.72	4.45	5.17
	Total	**1492611.02**	**1165278.31**	**2657889.33**

Source: https://apeda.gov.in/apedawebsite/organic/data.htm#Summary_Statistics_2021

State-wise Organic Farm Production for the Year 2020-21

Sl.No.	State Name	Organic Production (In MT)	Conversion Production (In MT)	Total Production (in MT)
1	Madhya Pradesh	1214919.50	177176.43	1392095.93
2	Maharashtra	752176.23	23598.76	775774.99
3	Karnataka	355718.73	0.00	355718.73
4	Rajasthan	237436.69	18949.46	256386.15
5	Uttar Pradesh	183089.90	319.14	183409.04
6	Odisha	128264.72	3587.28	131852.01
7	Gujarat	112797.25	5008.48	117805.73
8	Uttarakhand	46645.41	0.00	46645.41

Sl.No.	State Name	Organic Production (In MT)	Conversion Production (In MT)	Total Production (in MT)
9	Jammu and Kashmir	41043.93	0.00	41043.93
10	Kerala	27850.11	0.00	27850.11
11	Tamil Nadu	24068.86	0.00	24068.86
12	Meghalaya	21753.32	0.00	21753.32
13	Chhattisgarh	20630.65	0.00	20630.65
14	Andhra Pradesh	20145.27	0.00	20145.27
15	Assam	17839.16	0.00	17839.16
16	West Bengal	17434.79	0.00	17434.79
17	Himachal Pradesh	6368.45	0.00	6368.45
18	Haryana	5439.00	0.00	5439.00
19	Goa	3115.44	0.00	3115.44
20	Telangana	2509.68	0.00	2509.68
21	Sikkim	443.85	3.42	447.27
22	Tripura	348.53	0.00	348.53
23	Punjab	264.63	0.00	264.63
24	Manipur	27.74	0.00	27.74
25	Bihar	12.59	0.00	12.59
26	Nagaland	3.50	0.00	3.50
27	Arunachal Pradesh	1.09	0.00	1.09
	Total	**3240349.01**	**228642.97**	**3468991.98**

Source: https://apeda.gov.in/apedawebsite/organic/data.htm#Summary_Statistics_2021

Category-wise Production of Organic Commodities during Year 2020-21

Sl.No.	Category	Organic Production (In MT)
1	Oil Seeds	853754.86
2	Fibre	811007.77
3	Sugar	797627.60
4	Cereals and Millets	321006.27
5	Spices and Condiments	104820.81
6	PULSES	91039.63
7	Medicinal Plant Products	80533.52
8	Fresh Fruits and Vegetables	67350.21
9	Tea	42120.79
10	Coffee	22401.54
11	Flowers	13191.48
12	Dry Fruits	11499.82

Sl.No.	Category	Organic Production (In MT)
13	FODDER	11059.53
14	Others	5796.57
15	On farm Processed Food	4003.86
16	Tuber Products	3134.76
	Total	**3240349.01**

Source: https://apeda.gov.in/apedawebsite/organic/data.htm#Summary_Statistics_2021

1.6 Govt. Schemes, Promoting Organic Farming in India

The Government of India provides assistance for promoting organic farming across the country though different schemes.

1.6.1 Paramparagat Krishi Vikas Yojana (PKVY)

Paramparagat Krishi Vikas Yojana promotes cluster based organic farming with PGS (Participatory Guarantee System) certification. Cluster formation, training, certification and marketing are supported under the scheme. Assistance of Rs. 50,000 per ha/3 years is provided out of which 62 per cent (Rs. 31,000) is given as incentive to a farmer towards organic inputs.

1.6.2 Mission Organic Value Chain Development for North Eastern Region (MOVCDNER)

The scheme promotes third party certified organic farming of niche crops of north east region through Farmer Producer Organisations (FPOs) with focus on exports. Farmers are given assistance of Rs 25,000 per hectare for three years for organic inputs including organic manure and bio-fertilisers among other inputs. Support for formation of FPOs, capacity building, post-harvest infrastructure up to Rs 2 crore are also provided in the scheme.

1.6.3 Capital Investment Subsidy Scheme (CISS) under Soil Health Management Scheme

Under this scheme, 100 per cent assistance is provided to state government, government agencies for setting up of mechanised fruit and vegetable market waste, agro waste compost production unit up to a maximum limit of Rs 190 lakh per unit (3000 Total Per Annum TPA capacity). Similarly, for individuals and private agencies assistance up to 33 per cent of cost limit to Rs 63 lakh per unit as capital investment is provided.

1.6.4 National Mission on Oilseeds and Oil Palm (NMOOP)

Under the Mission, financial assistance at 50 per cent subsidy to the tune of Rs. 300 per hectare is being provided for different components including bio-fertilisers, supply of Rhizobium culture, Phosphate Solubilising Bacteria (PSB), Zinc Solubilising Bacteria (ZSB), Azatobacter, Mycorrhiza and vermi compost.

1.6.5 National Food Security Mission (NFSM)

Under NFSM, financial assistance is provided for promotion of bio-fertiliser (Rhizobium/PSB) at 50 per cent of the cost limited to Rs 300 per hectare.

As per international resource data from **Research Institute of Organic Agriculture** (FiBL) and the **International Federation of Organic Agriculture Movements** (IFOAM) Statistics 2020, India stands at 9th position in terms of certified agricultural land with 1.94 million hectare (2018-19).

Comparative Data in Regard to Organic Cultivation with other Countries

Sl.No.	*State*	*Position*	*Area under Organic Certification (in million ha.)*
1	China	3rd	3.14
2	USA	7th	2.02
3	India	9th	1.94
4	Brazil	12th	1.18

Promotion of natural farming under **Bharatiya Prakritik Krishi Padhati** (BPKP) of PKVY has been initiated to encourage use of natural on-farm inputs for chemical free farming. Andhra Pradesh and Kerala have taken up one lakh hectare and 0.8 lakh hectare area respectively for promotion of natural farming under BPKP. Similarly, continuous area certification and support for individual farmers for certification have also been initiated during 2020-21 to bring in default organic areas and willing individual farmers under the fold of organic farming

State agencies, **Primary Agricultural Credit Societies** (PACS), **Farmer Producer Organisations** (FPOs), entrepreneurs among others can avail loans for setting up of post-harvest infrastructure for value addition to organic produce under 1 lakh crore **Agriculture Infrastructure Fund** (AIF) of **Aatmanirbhar Bharat.**

1.7 Converting a Traditional Farm into an Organic Farming

If you have been doing traditional farming till now, you cannot suddenly jump to doing organic farming. The soil needs a transition period. The duration depends on the type of cultivation. Generally, it takes around 3 years for a traditional farm to convert fully into a certified organic farm.

1.8 National Programme for Oganic Production (NPOP)

Organic products are grown under a system of agriculture without the use of chemical fertilizers and pesticides with an environmentally and socially responsible approach. This is a method of farming that works at grass root level preserving the reproductive and regenerative capacity of the soil, good plant nutrition, and sound soil management, produces nutritious food rich in vitality which has resistance to diseases.

India is bestowed with lot of potential to produce all varieties of organic products due to its various agro climatic conditions. In several parts of the country, the inherited tradition of organic farming is an added advantage. This holds promise for the organic producers to tap the market which is growing steadily in the domestic and export sector.

As per the available statistics, India's rank 8th in terms of World's Organic Agricultural land and 1st in terms of total number of producers as per 2020 data (Source: FIBL and IFOAM Year Book, 2020).

The APEDA, Ministry of Commerce and Industries, Government of India is implementing the National Programme for Organic Production (NPOP). The programme involves the accreditation of Certification Bodies, standards for organic production, promotion of organic farming and marketing *etc.* The NPOP standards for production and accreditation system have been recognized by European Commission and Switzerland for unprocessed plant products as equivalent to their country standards. With these recognitions, Indian organic products duly certified by the accredited certification bodies of India are accepted by the importing countries. APEDA is also in the process of negotiation with South Korea, Taiwan, Canada, Japan *etc*

1.8.1 Area

As on 31st March 2021 total area under organic certification process (registered under National Programme for Organic Production) is 4339184.93 ha (2020-21). This includes 2657889.33 ha cultivable area and another 1681295.61 ha for wild harvest collection. Among all the states, Madhya Pradesh has covered largest area under organic certification followed by Rajasthan, Maharashtra, Chhattisgarh, Himachal Pradesh, Jammu and Kashmir and Karnataka. During 2016, Sikkim has achieved a remarkable distinction of converting its entire cultivable land (more than 75000 ha) under organic certification.

1.8.2 Production

India produced around 3496800.34 MT (2020-21) of certified organic products which includes all varieties of food products namely Oil Seeds, fibre, Sugar cane, Cereals and Millets, Cotton, Pulses, Aromatic and Medicinal Plants, Tea, Coffee, Fruits, Spices, Dry Fruits, Vegetables, Processed foods *etc.* The production is not limited to the edible sector but also produces organic cotton fiber, functional food products *etc.* Among different states Madhya Pradesh is the largest producer followed by Maharashtra, Karnataka, Rajasthan and Uttar Pradesh. In terms of commodities Oil seeds are the single largest category followed by Sugar crops, Cereals and Millets, Tea and Coffee, Fiber crops, fodder, Pulses, Medicinal/Herbal and Aromatic plants and Spices and Condiments.

1.8.3 Exports

The total volume of export during 2020-21 was **888179.68 MT.** The organic food export realization was around **INR 707849.52 Lakhs (1040.95 million USD).** Organic products are exported to USA, European Union, Canada, Great Britain, Korea Republic, Israel, Switzerland, Ecuador, Vietnam, Australia *etc.* In terms of export value realization Processed foods including soya meal (57 per cent) lead among the products followed by Oilseeds (9 per cent), Cereals and millets (7 per cent), Plantation crop products such as Tea and Coffee (6 per cent), Spices and condiments (5 per cent), Medicinal plants (5 per cent), Dry fruits (3 per cent), Sugar (3 per cent), and others.

1.9 Advantages of Organic Farming

- Organic farming in India is very economical, it uses no expensive fertilizers, pesticides, HYV seeds for the plantation of crops. It has no expenses.
- With the use of cheaper and local inputs, a farmer can earn a good return on investment. This is one of the most important benefits of organic farming in India.
- There is a huge demand for organic products in India and worldwide and can earn more income through export.
- Organic products are more nutritional, tasty, and good for health to chemical and fertilizer utilized products.
- Organic farming in India is very environment friendly, it does not use fertilizers and chemicals.
- These are some benefits of Organic Farming, which proves organic farming is profitable for everyone. We need to spread awareness about the advantages of organic farming in India to encourage Organic farming.

1.10 Disadvantages of Organic Farming

- Organic farming in India has fewer choices, and off-season crops are limited.
- Organic agricultural products are low in the early years. Farmers find it difficult to accommodate mass production.
- The main disadvantage of organic farming is the lack of marketing of the products and inadequate infrastructure.
- These are the information about Organic Farming in India with types of organic farming, methods, and advantages. We hope you enjoy this blog and get all the details regarding techniques used in organic farming.

Organic farming yields more nutritious and safe food. The popularity of organic food is growing dramatically as consumer seeks the organic foods that are thought to be healthier and safer. Thus, organic food perhaps ensures food safety from farm to plate. The organic farming process is more eco-friendly than conventional farming. Organic farming keeps soil healthy and maintains environment integrity thereby, promoting the health of consumers. Moreover, the organic produce market is now the fastest growing market all over the world including India. Organic agriculture promotes the health of consumers of a nation, the ecological health of a nation, and the economic growth of a nation by income generation holistically. India, at present, is the world's largest organic producers (Willer and Lernoud, 2019) and with this vision, we can conclude that encouraging organic farming in India can build a nutritionally, ecologically, and economically healthy nation in near future.

1.11 Comparison between Organic Farming and Conventional Farming

Differentiating Factor	*Conventional Farming*	*Organic Farming*
Before seeds are sown	Conventional Farmer fumigates his farm using harsh chemicals to exterminate any naturally existing pathogens or weeds and fertilizes the soil using petroleum-based fertilizers.	Organic farmer prepares his land and enriches the soil by sprinkling natural fertilizers like shellfishes, manures, or bone meals.
Before planting seeds	Conventional Farmer soaks the seeds in fungicides and pesticides to protect them from insects and pests. Chemicals are added to the irrigation water to prevent insects from damaging the planted seeds.	Organic farmer does not soak the seeds in chemicals nor irrigates them with water loaded with chemicals. He depends on natural rain and during dry months depends on stored rainwater.
To get rid of weeds	Conventional Farmer uses **weedicides** to remove the weeds.	Organic farmer physically removes the weeds out of the farm, although it would be very **labor-intensive**. Further, he could use a **flame weeder** to exterminate weeds or use animals to eat away the weeds.
During consumption	Any person consuming the products from the conventional farmers would absorb the pesticide and weedicide into the body, which could lead to **diseases** like cancer.	The absence of chemical application in OrF makes the food **safe** for consumption which is why people are going organic in record numbers today.

1.12 Challenges to Organic Farming in India

- ☆ **Lack of Awareness:** The most important challenge in the development of organic farming is the lack of awareness among farmers about the organic farming and its potential benefits.
- ☆ **Marketing Problems:** Before the beginning of organic crop cultivation, their marketability over conventional produce must be assured. Inability to obtain a premium price for the produce, at least during the nascent stages, leads to a setback.

All in all, organic agriculture is demanding and requires sound governmental frameworks such as a favourable legislation and support measures. Equally important is a good implementation of organic agriculture by farmers themselves. The challenges on the farm are manifold. They start with optimal farm and crop rotation planning during conversion and extend to daily decisions for optimising the production of individual crops in terms of costs and labour. The latter is important for increasing yields and profit margins across the crop rotation. Livestock challenges are similarly high. Organic farms strive for closing nutrient cycles, prioritising the use of on-farm or locally available resources. Fertile soil is the most important production capital of organic farmers. It must be built up through balanced crop rotation and organic fertilisation to ensure good yields and farm profitability. Figure 1.1 visualises the production system and the main challenges for organic farmer

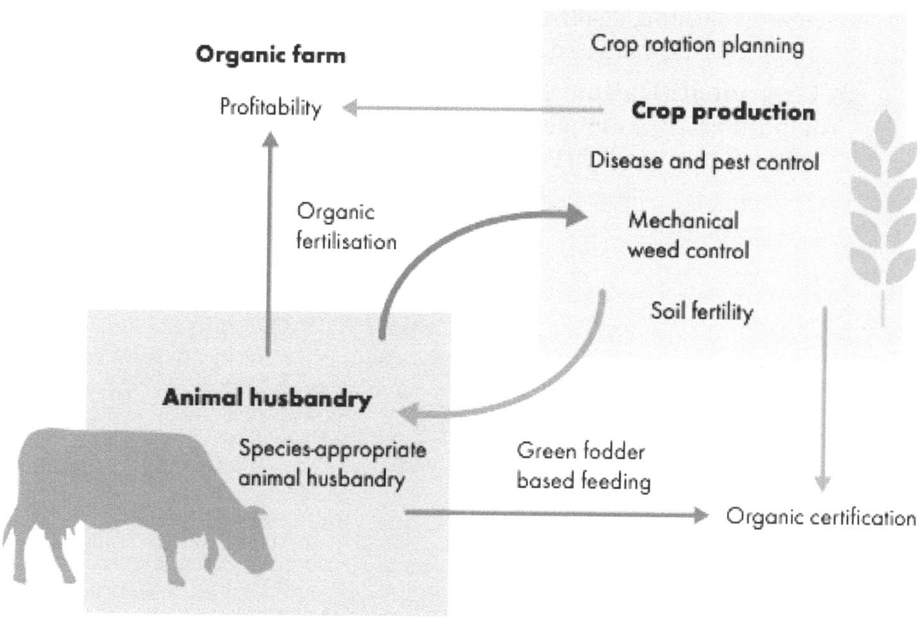

Figure 1.1. Most Important Challenges on Organic Farms.

- **Shortage of Biomass:** There could be a shortage of required quantities of the nutrients which could simply not be enough to meet the requirements.
- **Inadequate Infrastructure Support:** Despite NPOP adoption, the state governments are yet to formulate a credible mechanism and necessary policies for implementation. The certifying agencies are inadequate.
- **High Input Costs:** The costs of the organic inputs are more than those of industrially-produced agrochemicals used in the conventional farming system.

Crop production and animal husbandry are closely linked on organic farms. Coordination and efficient use of farm resources are important prerequisites for farm success.

1.13 Required Measures

- **Financial support**: Substantial financial support is necessary to promote organic farming. In India, organic farming does not receive the benefits of government subsidies.
- A program for OrF must be supported by full compensation both in cash and kind to the farmers in the event of a loss.
- **Market development:** It is a crucial factor to promote domestic sales. An important role of the government in this direction should be supporting the producer and consumer associations to market the products.
- **Awareness:** A vigorous campaign is crucial to highlight the benefits of organic farming against conventional farming and increase awareness among the farmers and consumers.
- **Crop identification:** Identification of crops for organic farming is important. For example, cotton growth in the rainfed areas or soybean cultivation in Madhya Pradesh could have great potential.

Chapter 2

Principles of Organic Farming

2.1 Introduction

Organic farming is very important because it produces naturally grown organic foods. Organic farming emphasizes soil fertility and maintains the soil naturally without using chemicals to help grow and protect crops. Now a day's organic farming is increasingly becoming more popular because people want less pesticides and fertilizers on our food. Organic farming not only helps us by providing better quality of food but also saves the environment by not spraying harmful pesticides into the air. Organic farming is essential for the safety of environment. It saves the planet for being polluted and damaged.

The International Federation of Organic Agricultural Movement (IFOAM) is an international body working since 1972 it regulates the organic farming.

The Four Principles of Organic Farming

- ☆ *Organic farming should sustain and enhance the health of soil, plants, animals and humans as one and indivisible.*
- ☆ *Organic farming should be based on the living ecological systems and cycles, work with them, emulate them and help sustain them.*
- ☆ *Organic agriculture should build on relationships that ensure fairness with regard to common environment and life processes.*
- ☆ *Organic farming should be managed in a precautionary and responsible manner to protect the health and well being of current and future generations and the environment.*

According to them "The role of organic agriculture, whether in farming, processing, distribution, or consumption, is to sustain and enhance the health of ecosystems and organisms from the smallest in the soil to human beings." Artificial practices like battery-raised animals, genetically modified foods and additives to animal feed do not include in organic farming.

Is Organic Farming Sustainable?

Organic agriculture supports the ecosystem balance and spares natural resources. This approach ensures the availability and productivity of lands in the future. It echoes with sustainable agriculture in this regard:

- Improves the soil quality;
- Reduces pollution with recycling and limitations of chemicals;
- Enhances the ecosystem biodiversity;
- Saves energy and resources, disapproves their excessive exploitation;
- Applies non-chemical methods of pest and weed management;
- Provides natural conditions to support livestock needs and behavior.

2.2 Principles of Organic Farming

There are four main principles of Organic farming according to IFOAM. Following are the principles on which organic farming is based on The principle of health, The principle of ecology, The principle of fairness and The principle of care The Four Principles of Organic Agriculture are as follows:

2.2.1 Principle of Health

Organic Agriculture should sustain and enhance the health of soil, plant, animal, human and planet as one and indivisible. This principle points out that the health of individuals and communities cannot be separated from the health of ecosystems - healthy soils produce healthy crops that foster the health of animals and people. The role of organic agriculture, whether in farming, processing, distribution, or consumption, is to sustain and enhance the health of ecosystems and organisms from the smallest in the soil to human beings. In particular, organic agriculture is intended to produce high quality, nutritious food that contributes to preventive health care and well-being. In view of this it should avoid the use of fertilizers, pesticides, animal drugs and food additives that may have adverse health effects.

2.2.2 Principle of Ecology

Organic Agriculture should be based on living ecological systems and cycles, work with them, emulate them and help sustain them. This principle roots organic agriculture within living ecological systems. It states that production is to be based on ecological processes, and recycling. Nourishment and well-being are achieved through the ecology of the specific production environment. For example, in the case of crops this is the living soil; for animals it is the farm ecosystem; for fish and marine organisms, the aquatic environment. Organic farming, pastoral and wild harvest systems should fit the cycles and ecological balances in nature. These cycles are universal but their operation is site-specific. Organic management must be adapted to local conditions, ecology, culture and scale. Inputs should be reduced by reuse, recycling and efficient management of materials and energy in order to maintain and improve environmental quality and conserve resources.

2.2.3 Principle of Fairness

Organic Agriculture should build on relationships that ensure fairness with regard to the common environment and life opportunities. This principle emphasizes that those involved in organic agriculture should conduct human relationships in a manner that ensures fairness at all levels and to all parties - farmers, workers, processors, distributors, traders and consumers. Organic agriculture should provide everyone involved with a good quality of life, and contribute to food sovereignty and reduction of poverty. It aims to produce a sufficient supply of good quality food and other products. This principle insists that animals should be provided with the conditions and opportunities of life that accord with their physiology, natural behaviour and well-being. Natural and environmental resources that are used for production and consumption should be managed in a way that is socially and ecologically just and should be held in trust for future generations. Fairness requires systems of production, distribution and trade that are open and equitable and account for real environmental and social costs.

2.2.4 Principle of Care

Organic Agriculture should be managed in a precautionary and responsible manner to protect the health and well-being of current and future generations and the environment. Organic agriculture is a living and dynamic system that responds to internal and external demands and conditions. Practitioners of organic agriculture can enhance efficiency and increase productivity, but this should not be at the risk of jeopardizing health and well-being. Consequently, new technologies need to be assessed and existing methods reviewed. Given the complete understanding of ecosystems and agriculture, care must be taken. This principle states that precaution and responsibility are the key concerns in management, development and technology choices in organic agriculture. Science is necessary to ensure that organic agriculture is healthy, safe and ecologically sound. However, scientific knowledge alone is not sufficient. Practical experience, accumulated wisdom and traditional and indigenous knowledge offer valid solutions, tested by time. Organic agriculture should prevent significant risks by adopting appropriate technologies and rejecting unpredictable ones, such as genetic engineering. Decisions should reflect the values and needs of all who might be affected, through transparent and participatory processes.

Chapter 3

Concept of Organic Farming

3.1 Introduction: An Overview

It is exclusively known for non-use of synthetic/artificial chemicals (fertilizer, pesticides, herbicides, growth hormones *etc.*). It is also known farming carbon because of heavy use of carbonaceous materials as a source of energy and plant nutrients. The approach and outlook towards agriculture and marketing of food has seen a quantum change worldwide over the last few decades. Whereas earlier the seasons and the climate of an area determined what would be grown and when, today it is the "market" that determines what it wants and what should be grown. The focus is now more on quantity and "outer" quality (appearance) rather than intrinsic or nutritional quality, also called "vitality".

This immense commercialisation of agriculture has also had a very negative effect on the environment. The use of pesticides has led to enormous levels of

Soil Organic farming is an agricultural approach that advocates healthy products free from components that may harm humans and nature. They include but are not limited to industrial pesticides, insecticides, fertilizers, clones, GMOs, chemical medications, hormones, growth-boosters, etc.The basic idea of organic agriculture is to provide food with optimum nutritional value and minimum dangerous ingredients, with only permitted substances used. The principle also requires 100 per cent natural forage for livestock and its further processing without synthetics. Organic farming practices involve care for field employees, and aim at maintaining the harmonious balance in the environment keeping it as alive and productive as possible.

chemical build up in our environment, in soil, water, air, in animals and even in our own bodies. Fertilisers have a short-term effect on productivity but a longer-term negative effect on the environment where they remain for years after leaching and running off, contaminating ground water and water bodies. The use of hybrid seeds and the practice of monoculture have led to a severe threat to local and indigenous varieties, whose germplasm can be lost forever.

In the name of growing more to feed the earth, we have taken the wrong road of unsustainability. The effects already show-farmers committing suicide in growing numbers with every passing year; the horrendous effects of pesticide sprays. The bigger picture that rarely makes news however is that millions of people are still underfed and where they do get enough to eat, the food they eat has the capability to eventually kill them.

Another negative effect of this trend has been on the fortunes of the farming communities worldwide. Despite this so-called increased productivity, farmers in practically every country around the world have seen a downturn in their fortune. The only beneficiaries of this new outlook towards food and agriculture seem to be the agro-chemical companies, seed companies and - though not related to the chemicalisation of agriculture, but equally part of the "big money syndrome" responsible for the farmers' troubles - the large, multi-national companies that trade in food, especially foodgrains.

Organic farming has the capability to take care of each of these problems. Besides the obvious immediate and positive effects organic or natural farming has on the environment and quality of food, it also greatly helps a farmer to become self-sufficient in his requirement for agro-inputs and reduce his costs.

In short, need of organic farming have been felt following reasons:

- ☆ The demand for organic food is steadily increasing both in the developed and developing countries with an annual average growth of 20-25 per cent.
- ☆ Decline in productivity of soil.
- ☆ Indiscriminate use of pesticides affects human and animal health, biodiversity, wildlife *etc.* and cause environmental pollution.
- ☆ High cost of inputs in conventional agriculture.
- ☆ Declining factor productivity.
- ☆ Deficiency of micronutrients.
- ☆ Global warming due to rise in carbon-dioxide and temperatures.
- ☆ Growth rate of agriculture production (1.5 per cent) is much below the population growth rate (2.0 per cent). Our country to be economically strong should improve on agriculture and allied enterprises.

3.2 Large Scale Farming

Today's chemical farms have little use for the skilled husbandry which was once the guiding principle of working the land. The emphasis today is solely on productivity - high input in exchange for high returns and productivity (mostly diminishing now however for farmers worldwide). Four important considerations - what happens to the land, the food it produces, the people who eat it and the communities which lose out - are overlooked.

The residue persistence in agricultural produce and global fertilizer consumption as shown below in Tables 3.1 and 3.2 respectively.

Table 3.1. Pesticide Residue Persistence in Agricultural Produce and Food Commodities

Commodity 2001 2002			
Samples (nos.)	*Contamination*	*Samples (Nos.)*	*Contamination*
Vegetables* (17 crops)	71261	529	63.5
	(12 per cent above MRL#)		(8.5 per cent above MRL)
Fruits** (12 crops)	38753	329	47.0
	(Less than MRL)		(approaches MRL)

* At Hisar all contaminated – 46 per cent above MRL, Heptachlor and Cypermethrin

** Fields in Faridabad – Vegetables, fruits, and flowers highly contaminated

Maximum residue limit (MRL)

Source: CCS Haryana Agricultural University (2003).

Table 3.2. Global Fertilizer Consumption (Mt nutrients)

N	*P_2O_5*	*K_2O*	*Total*
100.8	38.5	29.1	168.4
98.3	33.9	23.1	155.3
102.2	37.6	23.6	163 5
104.2	40.6	27.5	172.2
107.8	40.6	27.7	176.6
107.6	40.4	28.0	176.0

Source: IFA Agriculture, Dec 2013.

Table 3.3: Impacts of Processes on these different Life and Related Processes on Surface of the Soil

Process/Methods	*Impact/Fate of Processes*
Land exhaustion	The constant use of artificial fertilizer, together with a lack of crop rotation, reduces the soil's fertility year by year.
Fertilizers	High yield levels are produced by applying large quantities of artificial fertilizers, instead of by maintaining the natural fertility of the soil.
Nitrate run-off	About half of the nitrate in the artificial fertilizer used on crops is dissolved by rain. The dissolved nitrate runs off the fields to contaminate water courses.
Soil erosion	Where repeated deep ploughing is used to turn over the ground, heavy rains can carry away the topsoil and leave the ground useless for cultivation.
Soil compaction	Damage to the structure of soil by compression is a serious problem in areas that are intensively farmed. Conventional tillage may involve a tractor passing over the land six or seven times, and the wheeling's can cover up to 90 per cent of a field. Even a single tractor pass can compress the surface enough to reduce the porosity of the soil by 70 per cent, increasing surface run-off and, therefore, water erosion.
Agricultural fuel	As crop yields grow, so does the amount of fuel needed to produce them. European farmers now use an average of 12 tons to 1Km^2 of land.
Biocide sprays	The only controls used against weeds and pests are chemical ones. Most crops receive many doses of different chemicals before they are harvested.
Cruelty to animals	On most "modern" farms, all animals are crowded together indoors. Complex systems of machinery are needed to feed them, while constant medication is needed to prevent disease. The cruelty involved in managing, breeding, growing and slaughtering farm animals today is unimaginably repulsive and horrifying.
Animal slurry	With so many animals packed together in indoor pens, their manure accumulates at great speed. It is often poured into lagoons which leak into local watercourses, contaminating them with disease-causing organisms and contributing to algae-blooms.
Imported animal feed	Many farms are not self-sufficient in animal feed; instead they rely on feed brought into the farm. This often comes from countries which can ill afford to part with it.
Stubble burning	In countries where stubble is burned, large amounts of potentially useful organic matter disappear into the sky in clouds of polluting smoke
Loss of cultivated biodiversity	Large and other chemical farms tend to be monocultures growing the same crop and crop variety

Process/Methods	Impact/Fate of Processes
Threat to indigenous seeds animal breeds and species	Native cultivars and animal breeds lose out to exotic species and hybrids. Many native animal breeds are today threatened with extinction. The same holds true for many indigenous plant varieties which have disappeared within the space of one generation.
Habitat destruction	Agribusiness farming demands that anything which stands in the way of crop production is uprooted and destroyed. The wild animals and plants which were once a common sight around farms are deprived of their natural habitat and die out.
Contaminated food	Food, both plant and animal products, leaves the farm contaminated with the chemicals that were used to produce it.
Destruction of traditional knowledge systems and traditions	Rural indigenous knowledge and traditions, both agricultural and non-agricultural, is invariably connected to agriculture and agricultural systems.
Control of agriculture inputs and food distribution channel	The supply and trading in agricultural inputs and produce is in the hands of a few large corporations. This threatens food security, reducing the leverage and importance of the first and the last part of the supply chain - the farmer and the consumer.
Threat to individual farmers	Chemical agriculture is a threat to their livelihoods and changes their lifestyles, unfortunately not for the better.

3.3 Organic Versus Conventional

Organic farming is not a "new" concept; however, it was marginalized against the large-scale chemical based farming practices that have steadily dominated food production over the last 45 years. The difference between organic farming and modern conventional farming accounts for most of the controversy with claims and counter claims surrounding organic agriculture and organic food. Presently the comparison looks something like given below in Table 3.4.

Table 3.4. Comparison of Organic and Conventional Farming

Parameter	Organic Farming	Conventional Farming
Size	Smaller, marginal, dependent operations	Large scale, economically tied to major food corporation
Method	No use of purchased fertilizer and other inputs *e.g.* Pesticides, weedicides *etc.* less mechanization of the growing and harvesting process. Use of organic inputs like green manure, vermicompost, biofertilizers *etc.*	Heavy use of chemicals *e.g.* fertilization, use of pesticides *etc.* mechanized production using special equipment and facilities.
Technology	Nature based, environment friendly and sustainable.	Synthetic, harmful to environment and nutrient depleting.

Parameter	*Organic Farming*	*Conventional Farming*
Products	Good in taste, flavour, nutrition and free from chemicals.	Tasteless, less nutritious, may contain toxic residues of chemicals.
Market	Local, direct to consumer, on farm stands and farmers markets and through special wholesalers and retailer.	Wholesale with products distribution across large areas (average supermarket produce travels 100 to 1000 Km) and sold through high-volume.

3.4 Definitions

Organic farming is a reaction against large scale farming/exhaustive farming. A large number of terms are used as an alternative to organic farming. These are: biological agriculture, ecological agriculture, do-nothing agriculture, bio-dynamic, organic-biological agriculture and natural agriculture.

The essential concept of these practices remains the same, *i.e.*, back to nature, where the philosophy is to feed the soil rather than the crops to maintain soil health and it is a means of giving back to the nature what has been taken from it (Funtilana, 1990).

According to the *Codex Alimentarius Commission*,'organic agriculture is a holistic production management system that avoids use of synthetic fertilizers, pesticides and genetically modified organisms, minimizes pollution of air, soil and water, and optimizes the health and productivity of interdependent communities of plants, animals and people'.

- **Pure Organic Farming:** This accounts complete exclusion of inorganic fertilizers and pesticides, but advocates the use of organic manures and biological pest control methods.
- **Integrated Green Revolution Farming:** Under this option, the basic trends of the green revolution such as intensive use of external inputs, increased irrigation, development of high yielding and hybrid varieties as well as mechanizations of labour are retained with much greater efficiency on the use of these inputs with limited damage to the environment and human health. For this purpose some organic techniques are developed and combined with the high input technology in order to create Integrated Systems such as, "Integrated Nutrient Management" (INM), "Integrated Pest Management" (IPM) and biological control methods which reduce the need for chemicals.
- **Integrated Farming System:** This option involves low input organic farming in which the farmers have to depend on local resources and ecological processes, recycling of agricultural wastes and crop residues.
- **Biodynamic farming**: Biodynamic farming is done through the steps given below:

- To apply zodiac principles to the process of cosmic integration; that is to say, the moon constellations in seed sowing and harvesting of crops.
- To apply biodynamic methods for compost preparation and field sprays.
- Use of vermicompost, vermi-wash.
- Use of bone meal and wood ash.
- Use of plant-extract, cakes to manage soil and plant pests.
- Use of cow urine mixed with some plant chemicals, such as menthol or alone to control plant diseases and pests.
- *Equisetum arvense* has been used to control fungal diseases successfully.
- *Melia azadirach* (Drek,) is used like Neem to control pests in the form an extract.
- Plants like *Lantana camara* can be used to control insect pests or as a mosquito repellent.
- Use of cow pat pit or liquid manure or compost heap methods provide sufficient nutrients to the growing crops

3.5 Components of Organic Farming

There are assumptions throughout the organic literature of differences between organic and conventional systems with respect to their effects on soil physical properties, soil insect fauna and nutrient flow within the soil, crop health and nutritional value of the harvested crop. Different components of organic farming are as follows:

3.5.1 Crop and Soil Management

Organic farming system encourages the use of rotations and measures to maintain soil fertility. Carefully managed soil with a high production of humus offer essential advantages with respect to water retention ion exchange, soil erosion and animal life in the soil. Green manuring and inter-cropping of legumes is another important aspect for biological farming systems not only in regard to weed control but also in reducing the leaching of nutrients and in reducing soil erosion. A green cover throughout most of the year is one of the main goals of such farming methods. Depending on the green manure mixture or the legumes used for under sowing, there may be an increased soil organic matter and soil N_2 as well as in other nutrients.

3.5.2 On-farm Waste Recycling

Increase price of chemical fertilizers have enables organic wastes to regain an important role in the fertilizer practices on the farm. Good manure management means improved fertilizers value of manure and slurry and less nutrient losses. Composting of all organic wastes in general and of Farm Yard Manure (FYM) or feed-lot manure in particular is important in organic farming.

3.5.3 Non-chemical Weed Management

Weed management is one of the main concerns in organic agriculture. Generally, all aspects of arable crop production play an important role in a system approach to problems. The elements to consider in preventing weed problems are crop rotation, green manuring, manure management and tillage. Mulching on a large scale by using manure spreaders may also be useful in weed control.

3.5.4 Domestic and Industrial Waste Recycling

Sewage and sludge use for crop production can form an important component of organic farming if treatment and application methods are improved further.

3.5.5 Energy Use

In the energy requirement for production measured per rupees of produce for organic farms is only one third of what it is for their conventional counterparts. Because N-fertilizer and pesticides are not used by biological farmers, the comparison of total energy input/ha with total energy output favors biological farming systems.

3.5.6 Food Quality

Food quality is one of the main issues, which concerns both scientists and consumers. Nitrates in water and farm produce, desirable components, pesticides residues, keeping quality and physiological imbalances are some of the important aspects of food quality.

3.5.7 Ecological Agriculture

The growing concern about environmental degradation, dwindling natural resources and urgency to meet the food needs of the increasing population are compelling farm scientist and policy makers to seriously examine alternative to chemical agriculture. As reported by Vankataramani (1995) case studies shows that when chemical farm incurred about 11.250 towards the cost of cultivation of rice. An organic farm spend rupees 10,590 to produce 5625 kg paddy and 8 tonnes of straw/ha. The net returns from the ecological farming system at the current cost of rupees 3.34/kg paddy is rupees 8,197.50. In chemical farming, the net profit is rupees 7500. If one gets a premium price for the poison force, organically grown rice, the economic returns from the ecological farming system will highly encouraging.

3.5.8 Integrated Intensive Farming System (IIFS)

IIFS involves intensive use of farm resources. To be ecologically sustainable, such intensification should be based on techniques which are knowledge intensive and which replace to the extent possible, market purchased chemical inputs with farm grown biological inputs.

3.5.9 Value of Organic Farming

The value of organic materials as fertilizers and soil conditions is often misunderstood and has been the source of some controversy. The simplest and the most common means of estimating the value of organic amendments is by assigning the market value of the potentially available plant nutrients, they contain nitrogen-phosphorus-potash (NPK) and other micro-organisms.

Chapter 4

Certification, Labelling and Accreditation Process

4.1 Organic Certification

It is a certification process for producers of organic food, seeds and other organic agricultural products. In general, any business directly involved in food production can be certified, including seed suppliers, farmers, food processors, retailers and restaurants. Organic seeds in general are double certified as they are certified twice, once by the seed certification team and again by the organic certification and monitoring team.

Requirements for certification and seed demand vary from country to country, and usually involve a set of production standards for growing, storage, processing, packaging and shipping of organic produce that include:

- Avoidance of synthetic chemical inputs (*e.g.* fertilizer, pesticides, antibiotics, food additives, *etc*) and genetically modified organisms and seeds
- Use of farmland that has been free from chemicals for a number of years, if not at least for past three years (Conversion period)
- Maintaining strict physical separation of organic seed fields from inorganic seed fields is a mandate (Buffer zone)
- Undergoing periodic on-site inspections

4.2 Purpose of Organic Certification

As of now there is no organic seed certification system in India but there are separate bodies that do organic certification and seed certification. This organic certification addresses the need of growing worldwide demand for organic food. It is intended to assure quality and prevent fraud. For organic producers, certification identifies suppliers of products approved for use in certified operations. Certification is written assurance given by Certification board that an identified production and processing system is methodically assessed and confirms to meet the specific requirement of organic standards. In India all the organic products meet the standards prescribed by the NPOP.

4.3 Certification Process

In order to certify a farm, the farmer is typically required to engage in a number of new activities, in addition to normal farming operations: Study the organic standards, which cover in specific detail what is and is not allowed for every aspect of farming, including storage, transport and sale.

4.3.1 Steps in Organic Certification

1. **Receipt of application form:** the farmer intending to get his farm organically certified must apply to the organic certification board. The documents required for applying are duplicate of application form, Pan card, Annual cropping pattern, Field map, General details of the farm, Soil and water analysis report, Chitta (land documents), a written annual production plan must be submitted, detailing everything from seed to sale (seed sources, field and crop locations, fertilization and pest control activities, harvest methods, storage locations, *etc.*)
2. **Scrutiny and registration of application:** the application received along with the others farm or field details are verified by the inspector if it meets the requirements then the application is forwarded for the registration. For the registration the farmer must pay a prescribed amount. Once the farm is registered it must be strictly maintained under the organic conditions only.
3. **Inspection and evaluation of the farms and documents:** annual on-farm inspections are required, with a physical tour, examination of records, and an oral interview is done. Record keeping written, day-to-day farming and marketing records maintained and will be checked, the farmers must have covered all activities like biodiversity conservation and buffer zone. The farmer must be available for inspection at any time. In addition, short notice or surprise inspections can be done by the certification officer.
4. **Sampling of soil, water and plant products if necessary:** if the certification officers doubt that the grower has done a mal-practice then he has all the right to collect the soil, water and plant sample. Analysis of

plant and soil sample will be done and if the results indicate the presence of any chemicals or toxic substance then their certificate will be taken back.

5. **Issue of certificate to eligible organic farms:** if the grower has maintained his farm purely under organic condition then a certificate will be given to him assuring others that he is an organic grower. The certificate is online generated, and it takes around six months from date of application.

4.4 Organic Standards

- Conversion period it is the time required for the conversion of inorganic field to completely organic field. Its two years for annuls and three years for perennials.
- Buffer zone: Area of three-meter square must be left in border of field to separate the organic field from inorganic field. If the organic field is in low regions, then there must be a drench dug to avoid entry of polluted water in the form of runoff to organic field.
- One per cent of area must be covered under the trees to protect the biodiversity and the farm must have a live fencing.
- Selection of crop and variety: the crop must be suitable to the area and season. The seeds used must be organic if such seeds are not available then the farmers can use the commercial seed in the first year and from the consequent year he must use the seeds that are produced in his farm.
- The crops grown must be divers as it avoids loss to grower.
- Only biodegradable material of microbial, plant or animal origin should be used, minimize nutrient loss is needed and accumulation of heavy metals and other pollutants must be prevented.
- Origin farm must encourage the soil and water conservation.
- Weeds should not be controlled chemical method. 9. No use of synthetic growth regulators.

Registration can be done by

1. **Individual farmers:** The farmer willing to register a field under organic must be a legal owner of the land and there no limit of area for certification.
2. **Group:** All the members of the group must be within the revenue district and no restriction on the area for certification but numbers of farmers with more than 10 ac of land should be less than 50 per cent of total area of group.
3. **Corporate bodies:** Group of farmers can come together and register as a corporate body. They can produce organically, process and sell but there should be an office setup to monitor the activities.

4.5 Certification and Product Labelling

Being able to put the word "organic" on a food product is a valuable marketing advantage in today's consumer market as people have become more health conscious. Certification is intended to protect consumers from misuse of the term organic, and make buying of organic products an easy task. However, the organic labelling made possible by certification.

In India the following certificates and labels are mandate to produce, sell and export any organic produce.

- ☆ **Scope certificate:** This is the certificate that is provided by the organic certification agency which ensures that field is maintained organically according to the NPOP standards. The certificate needs to be renewed every year by paying a prescribed fee. Scope certificate contains the following details of the organic grower and the field.
- ☆ **Transaction Certificate:** Transaction Certification is an authorization certificate given by APEDA stating that the products or stocks mentioned in transaction certificate has been produced and/or prepared in compliance with the current NSOP/NPOP standards. This certificate shall also be issued to the seller, if the seller provides the required details after the verification by the Organic Certification Inspector and evaluator.

4.6 Certification Process

Third party certification process starts with the adoption of National Standard of Organic Production (NSOP) on farm followed by the registration of production unit with one of the accredited certification body. There are 29 accredited certification bodies for granting organic certification. Producers can choose any one of them for their farm certification.

4.6.1 Scope Categories for Certification

NPOP Certification is available for following scope categories:

a. Crop production and wild harvest,
b. Livestock, Sericulture
c. Apiculture
d. Aquaculture
e. Organic Food processing and Handling
f. Organic Animal Feed Processing and Handling
g. Organic Mushroom
h. Seaweeds, Aquatic Plants and Greenhouse Crop Production

In order to certify a farm operation, the producer is typically required to engage in a number of new activities, in addition to normal farming operations which includes:

- Study the organic standards, which cover in specific detail what is and what is not allowed for every aspect of farming including storage, transport and sale.
- Compliance: Farm facilities and production methods must comply with the standards which may involve modifying facilities, sourcing and changing suppliers, *etc.*
- Documentation: Adequate documentation detailing farm history, current set-up, operational activities including details on inputs used, farming operations, facilities in use, source of contamination and methodologies implemented to prevent contamination, test reports on soil, water, plant, products *etc.*
- Planning: A written annual production plan must be submitted, detailing everything from seed/breed to sale: seed/breed sources, field and crop locations, fertilization and pest control activities, harvest methods, storage locations, livestock breeds, housing management, grazing, feed supplements, ensuring welfare and natural rearing practices and ensuring comfort at all stages for animals *etc.*
- Inspection: Annual on-farm inspections are required with a physical inspection, examination of records and an oral interview.
- Fee: A fee is to be paid by the operator to the certification body for annual surveillance and for facilitating a mark, which is acceptable in the market as symbol of quality.
- Record keeping: Written, day-to-day farming and marketing records, covering all activities must be available for inspection at any time.

In addition, short-notice or surprise inspections can be made and specific tests (*e.g.* soil, water, plant/animal product) may be requested. For first-time certification, the soil/livestock/animals must meet basic requirements of being free from use of prohibited substances (synthetic chemicals, *etc*) for a number of years. A conventional farm must adhere to organic standards for this period, often, two-three years. This is known as being in transition or conversion. Transitional crops/livestock products are not considered fully organic.

Certification for processing and handling operations other than farms is similar. The focus is on ingredients and other inputs, and processing and handling conditions. A slaughterhouse and milk handling facilities are required to follow the standard requirements for facilities and operations including equipments. A transport company would be required to detail the use and maintenance of its vehicles, storage facilities, containers, and so forth. A processing unit would have its premises inspected and its suppliers verified as certified organic.

4.6.2 Certification Procedure in Brief

- Application is made to the certification agency in the prescribed format with necessary farm and process details
- Screening of application by certification agency and if necessary further details/clarification sought
- Cost estimate comprising of certification charge, inspection charge, travel cost, reporting cost, laboratory charges *etc* is sent for acceptance
- Acceptance of cost by the grower/producer
- Signing of agreement between grower/producer and certification agency
- Certification agency seeks cropping/production/cultivation/processing plan and supply a copy of the standards to the grower/producer to follow
- Certification agency raises an invoice and asks the producer payment of initial fee
- Grower/producer pays the fee
- Inspection schedule is worked out
- Inspection is carried out at one or more than one occasion
- If required unannounced inspection can also be done. In case of doubt, the inspection team can also draw plant/soil/raw material/input/product sample for laboratory analysis.
- Inspection report/(s) submitted to the certification committee
- Certification agency asks for final payment
- Final payment is made
- Certification is granted
- On grant of scope certificate Producer/operator applies for license for use of India Organic Logo
- Certification body grants the license for use of India Organic Logo
- Grower/producer releases the stock for sale with Certification Mark

India Organic Logo

4.6.3 Inspection Procedure

Certification Bodies employ standard inspection procedures as per ISO19011 comprising of following:

i. A qualified and trained inspector is assigned to inspect the operations of the operator having adequate competence and no conflict of interest

ii. The same inspector shall not visit the same operator more than two years in a row.

iii. Operators have neither the right to choose nor to recommend inspectors.

iv. The operators have the right to be informed about the identity of the inspector before the inspection visit and to raise objections related to any potential conflict of interest.

v. Sufficient information is made available to the inspectors about the operator to allow proper preparation by the inspector which includes earlier inspection findings, a description of activities/processes, maps/plans, product specifications, inputs used, earlier irregularities, infringements, conditions and disciplinary measures.

vi. The checklists used during the inspection and the reports emanating from the inspection shall be comprehensive covering all relevant aspects of the production standards and shall adequately validate the information provided.

vii. The inspector shall have access to all relevant facilities including accounts and other documentation of the operator. Certification Bodies shall have access to any non-organic production unit or units associated by ownership or management.

viii. The inspector shall take precautionary measures by assessing the risk of noncompliance during the inspection. When an irregularity is committed by the operator relating to organic production as non-compliance of NPOP, the entire lot or production affected by irregularity shall be made to be removed from the production site/chain and sanctions shall be imposed on the operator. APEDA shall be informed within 30 days about the action taken on the operator.

ix. Inspection checklist, reports and inspection shall follow specified methods to facilitate a non-discriminatory and objective inspection procedure.

x. Reports shall be designed to allow for elaboration and analysis by the inspector on areas where compliance might be partial; standards might not be clear *etc.*

xi. Inspection reports shall give adequate information on what was actually checked including but not restricted to:

- Date and time of inspection
- Persons interviewed

- Crops/products requested for certification
- Fields and facilities visited
- Documents reviewed
- Buffer zones
- Risk of drift
- Risk of contamination
- Inspector's observations
- Calculation of input/output norms, production estimates *etc.*
- Assessment of production system of operator
- Assessment of the use of logos/approvals (India organic logo, product logo as well as the Certification Body's logo)
- Product reconciliation and verification of stock
- Interview with responsible persons
- Evaluation of compliance to standards and Certification requirements.

The standard procedures for inspection and certification of production at the production farms (individual and grower groups), wild collection, processing units (including sub contracted units) and at all stages in handling (storage units, packaging, shipments *etc*) includes:

a) Inspection of parallel production of farms,
b) Inspection of Processing Units
c) Inspection of grower groups
d) Inspection of wild production collection
e) Inspection of all stages in handling
f) Inspection of Packed Products
g) Inspection of Storage Facilities
h) Inspection of Transport Facilities
i) Inspection of Chain of Custody
j) Inspection for detection of use of Genetically Engineered/modified Products

a) Inspection of Parallel Production of Farms

i) Buffer zones are maintained for demarcation
ii) Crops are visually distinguishable
iii) Inspections are carried out at critical times
iv) Inspection is done in a timely manner.
v) Accurate production estimates are available

vi) The crops are harvested in such a way that there are reliable methods to verify the actual harvest of the respective crops.

vii) Appropriate storage capacity exists to ensure separate handling

viii) The documentation regarding the production is well managed and makes a clear distinction between certified and non certified production.

b) Inspection of Processing Units

During the inspection of the processing units, the following shall be taken care:

i) The Inspector shall verify that sufficient quantities of organic ingredients are used and that organic integrity is maintained through all stages of processing.

ii) The inspector shall review all ingredients and their sources to ensure that ingredients meet organic standards.

iii) The inspector shall also review product formulation to determine if they meet labeling standards

iv) The inspector shall verify the existing record keeping system and evaluate whether it is adequate for tracking organic products.

v) The inspector shall conduct an audit trail to track the product from the receipt of raw material/ingredients, ingredient storage through all stages of processing, packaging, labeling, warehousing, shipping and sales of the finished product.

vi) The inspector shall conduct a sample audit review which consists of randomly choosing a finished product(s) either a sales invoice, a product purchased or product seen in the warehouse. The inspector shall record the Lot number on the finished product and follow the product back to the receipt of incoming ingredients on the record keeping system. The inspector shall inspect all the subcontracted units annually.

c) Inspection of Grower Groups

i) The external inspection by the Certification Body shall be planned after internal inspections of all the farmers are carried out by the Internal Control System (ICS) twice annually

ii) The Certification Body shall have a standardized format for sourcing the information from the grower groups which shall include list of farmers, location on an area map, year of joining in the grower group, date of internal inspections, area of cultivation, crops and yield estimates. iii) The inspector shall verify that new farmers are included in the group only after the internal inspections are completed.

iv) The inspector shall carry out the risk assessment of the ICS and the inspector shall draw a sample of farms for visiting the farmers in the ICS.

v) The inspector shall prepare a list of farms having 4 Hectare or more and all such farms having 4 Hectare or more shall be inspected separately. The 4 Hectare and above farms shall not be included in the sample of farmers drawn for re-inspection.

vi) The inspection shall include a witness audit of the internal inspector for assessing his knowledge on inspection procedures.

vii) The inspector shall verify the documentation of the ICS that adequate records of inspections are maintained Instances of non-compliance and the active measures taken by the ICS with special reference to sanctions shall be assessed from the documentation.

viii) Internal control records are in compliance with the findings of the Certification Body's sample inspection results.

ix) The inspector shall interview the farmers, ICS manager to assess the knowledge of operator on NPOP standards.

x) The inspector shall verify the collected information from the ICS with the submitted information by the grower group during registration/renewal.

d) Inspection of Wild Production and Collection

i) To verify that the area of collection is properly identified on appropriate maps issued by the concerned Government Authorities. The map shall be large and distinct enough to reduce the risk of mixing up with non-certified production. However, wherever community rights are recognized under Forest Rights Act, 2006, Gram Sabha letter can be considered for verification of collection area by the community.

ii) Verification of operator records of all collectors and the quantities bought from each collector.

iii) Visit to an appropriate portion of the certified area.

iv) Visits and interviews of the concerned in the supply chain such as collectors, local agents, landowners and other parties (environment agencies, NGOs *etc.*)

e) Inspection of all Stages in Handling

Each step in the handling of a product shall be inspected at least once annually (storage units, packaging, shipment *etc*). Any person who sells a product (raises invoice) shall be registered and certified which applies until the product is in its final package/has its final label.

f) Inspection of Packed Products

The accredited Certification Bodies are not obliged to have a system for inspection of products that are further handled after being packed in the final consumer package, and/or after issuing of a transaction certificate. The accredited

Certification Bodies however, are obliged to take action where there is reason to believe that the standards have been or may be violated at such later stages.

g) Inspection of Storage Facilities

Depending on the kind of storage, the product, packing, prevailing storage practices (*i.e.* fumigation) and the time of storage, inspections shall be required. Accredited Certification Bodies shall conduct a risk assessment to determine future need for inspection of all storage facilities including port facilities.

h) Inspection of Transport Facilities

Transport is not certified as such, but remains under the responsibility of the operator owning the product during the transport.

i) Inspection of Chain of Custody

Accredited Certification Body shall not issue any license to use its certification mark or issue any certificate for any products unless it is assured of the chain of custody of the product where steps in the production chain have been certified by other accredited Certification Bodies under NPOP as per the National Standards of Production.

j) Inspection for Detection of Use of Genetically Engineered Products

Accredited Certification Bodies shall implement a system of inspection for potential use of genetically engineered products. When use of such products is detected at any stage, certification shall not be granted. When there is a risk of contamination of genetically engineered products, the following samples shall be tested in identified APEDA approved laboratories.

- Seeds and planting stock
- Production inputs
- Livestock feed
- Processing aids
- Ingredients

4.6.4 Inspection Methods and Frequency

Inspection methods and frequency of inspection shall be determined by the certification body, based upon following factors:

- Intensity of production
- Type of production
- Size of operation
- Outcome of previous inspections and the operator's record of compliance
- Any complaints received by the programme
- Whether the unit or operator is engaged only in certified production

- ☆ Contamination and drift risk
- ☆ Complexity of production, based upon following factors:

Announced Annual Inspections

(i) Inspection of certified operators shall take place at least once annually.

(ii) Inspection of sub-contracted operators or units shall take place at least once annually.

(iii) Timing of inspections shall not be so regular as to become predictable.

(iv) There shall be provisions for more inspections with respect to the factors stated below.

Unannounced Inspections

(i) The selection of operators for unannounced inspection shall be based on risk analysis carried out by the Certification Body annually.

(ii) A minimum of 10 per cent of unannounced inspections to be carried out annually by the Certification Bodies.

Risk Assessment

(i) Risk assessment is done covering all scope of activities

(ii) The risk assessment procedure cover the criteria for determining the risk category as high, medium or low

(iii) Based on the procedure of risk assessment, 10 per cent inspections are required to be carried out by the Certification Body annually in addition to the unannounced inspections

4.7 Grower Group Certification

1. Growers Groups are organized group of farmers/producers who intend to produce organic products/engage in organic processes in accordance with the National Standards of Organic Production (NSOP).

   The grower group shall be based on the Internal Control System (ICS) and shall apply to grower groups, farmers' cooperatives, contract production and small scale processing units. The producers in the group must apply similar production systems and the farms should be in geographical proximity. Individual farms with land holding of 4 ha (10 acres) and above can also be a part of the group but will have to be inspected separately every year by the accredited Certification Body. The total area of such farms shall be less than 50 per cent of the total area of the group. The grower group shall consist of minimum 25 and maximum 500 farmers. Processors and exporters/traders can own/manage the Internal Control System (ICS) but will have to be inspected annually by the external Certification Body. Separate certificates (Scope and Transaction

Certificates) are required to be issued for the ICS, processors and traders to maintain the traceability of the product flow.

The Certification Body shall not certify if there is no ICS as per NPOP and 100 per cent internal inspections are not conducted. In case the farmer group does not maintain an Internal Quality System as described, the Certification Body shall inspect all the Individual farms.

2. **Constitution of ICS -** The ICS will have a registered legal identity and have a constitution of the organization and shall be presented by an organizational chart. For implementation of the procedures to maintain the internal control system, responsibilities shall be delegated to individual members/committees for carrying out specific activities
3. **Internal Control System:** Group Certification is based on the concept of Internal Quality Management System which comprises of following:
   - Implementation of the Internal Quality System
   - Internal standard
   - Risk assessment
4. **How to Develop an ICS** - The following are minimum requirements for setting up an ICS for grower groups:
   - Development of Internal Control System (ICS) manual containing policies and procedures
   - Identification of farmers in the group
   - Creation of awareness about Grower Group Certification
   - Identification of qualified/experienced personnel for maintaining the Internal Control System
   - Give necessary training in production and ICS development
   - Implementation of the policies and procedures
   - Review and improvement of the ICS document for maintaining a harmonized quality management system
5. **Internal Standards:** Internal Standards shall be prepared in local language by the IQS Manager for the region of operations under the framework of NPOP standards. If the farmers are illiterate, the Internal Standards shall contain illustrations in the text for better understanding. The internal standards would contain:
   - Definition of production unit
   - How to deal with part conversion
   - Conversion period
   - Farm production norms for the entire production unit.
   - Harvest and post harvest procedures.

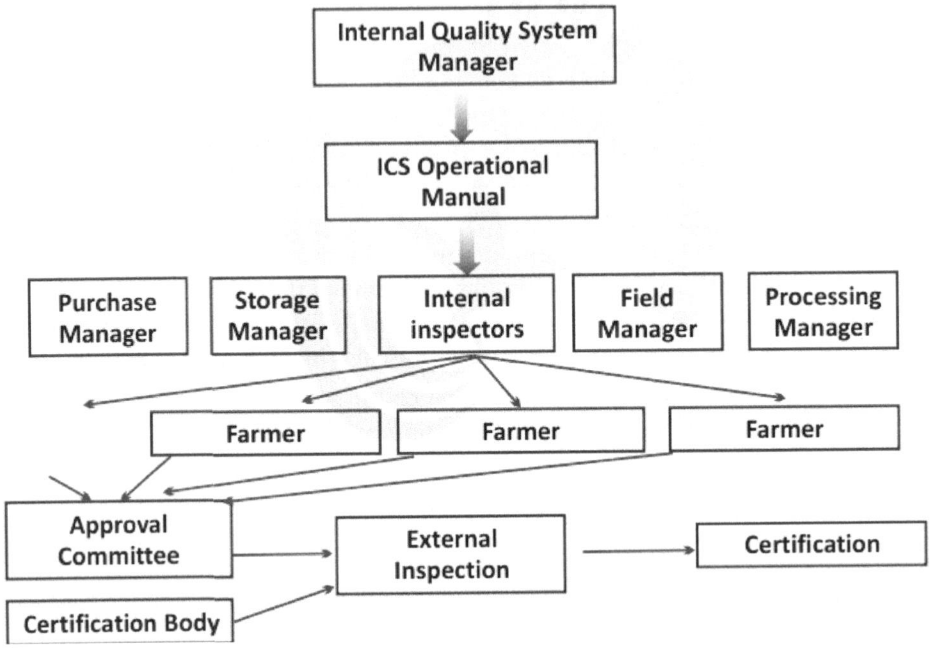

Typical Operational Structure of ICS.

6. **Conflict of Interest -** The ICS personnel shall not have any conflict of interest that might hinder the work. All possible conflicts shall be declared in a written statement.
7 **Scope of certification -** The certification shall be granted by the accredited Certification Body to the group as per NPOP.
8. **Trade -** The group will market the products under a single entity. For trading the products from the group of producers, the ICS shall draw up relevant procedures
9. **Procedures for Implementation of Internal Control System**
   (i) Registration of members: All members of the group will be legally registered under a single entity (name) with address of its operations (location, taluka, village)
   (ii) Provision of documents to the members of the grower group
   - ✰ Copy of ICS Manual
   - ✰ Internal standards document in local language
   - ✰ NPOP document
   - ✰ Definition of the production unit
   - ✰ Field records may be included in internal farm checklist
   - ✰ Farm Entrance Form including last use of prohibited substances

- ☆ Prevailing farming system and package of practices available for the area
- ☆ Details and description of the various steps required for the process flow, right from cultivation to harvest and sales of the products.
- ☆ Written contract of the grower with the group.
- ☆ Annual farm inspection checklist
- ☆ Information on training programmes and provision of advisory services by the field officers.

10. **Operating Document:** The ICS manager shall prepare the operating document which will contain the following: -
    - ☆ An overview map (village or community map) showing location of each member's production unit. The map should indicate the crops cultivated in rotation and also mark any farm in an area, which could be identified as high risk due to drift from non-conventional farms.
    - ☆ Farmer's list with code and name of the farmer, total area, area under crop, date of registration with the group, date of last use of forbidden products, date of internal inspection, name of internal inspector, result of internal inspection (separate lists for in-conversion farmers)
    - ☆ List of sanctioned farmers with the reason and the duration of this sanction (if relevant).
    - ☆ The risk shall be assessed by IQS manager for the grower group every year. The risk assessment should be made at the farm level, processing, transporting and during trade. The ICS will take all measures to minimize the identified relevant risks.

    Critical control points for risk assessment

    - ☆ Measures taken by the farmers to deal with part conversion (if farmers still grow some non-organic crops).
    - ☆ Conversion period
    - ☆ Production rules for the whole production unit e.g seeds, fertilization and soil management, pest management, approved inputs, preventions of drifts, animal husbandry.
    - ☆ Harvest and post harvest procedures.
    - ☆ Processing and handling procedures

11. **Internal Inspections**
    - ☆ At least two inspections of the group (one in growing season of each crop) shall be carried out by the internal inspector and will be documented. The inspection will be carried out in presence of the member or his representative and must include a visit of the whole farm, storage of inputs, harvested products, post harvest handling and animal husbandry.

- The internal inspector will also verify if the internal standards have been followed and whether the conditions of the previous internal inspection have been fulfilled.
- The visit of the internal inspector will be documented in the farm inspection checklist duly signed by the inspector and countersigned by the member or his representative.
- In case of severe non-compliance, the results will be reported immediately to the ICS manager and all measures will be taken according to the internal sanction procedures.

12. **Internal Approval -** The ICS manager will have a defined procedure to approve or impose sanction on the farmers in the group. All internal farm checklists are screened by internal approval staff with special focus on the critical control points of risk/difficult cases.

13. **External Inspections**

    a) The accredited Certification Bodies shall undertake inspections of the ICS after ensuring that 100 per cent internal inspections by the ICS have been undertaken at least twice for all the registered members of the grower group. The accredited Certification Body will inspect some of the farms for the evaluation of the grower group for efficient internal control system for compliance with the NPOP Standards. The accredited Certification Body shall inspect farms of 4 Hectare and above separately in addition to the sample of farms having the area of less than 4 Hectare. The sampling plan for inspection shall be based on the inspector's perception of risk based on the following factors:

    - Size of holding
    - Number of the members in the group
    - Degree of similarity between the production system and crop system
    - Inter-mingling/contamination
    - Parallel production
    - Split production
    - Local hazards
    - Change in the production plan
    - Joining of new members in the group

    b) **Risk Assessment:** A minimum sample size of the members in the grower group shall be inspected by the accredited Certification Body. The sample size shall be determined as square root of the number of members registered in the grower group. Based on the risk assessment made by the accredited Certification Body prior to the inspection visit, the number of sample inspections shall be planned. The risk assessed by the accredited Certification Body shall be documented. The accredited Certification Body shall establish criteria for assessment risk under high, medium and low categories. The accredited Certification Body shall follow the given pattern below for minimum number of farmers for inspections:

- High risk: 2 X square root of number of farmers
- Medium risk: 1.5 X square root of number of farmers
- Low risk: square root of number of farmers

14. **Yield Estimates -** Yields will be estimated for each crop for individual farmer in the group by the ICS and counter-checked with the estimates during external inspection by the accredited Certification Body
15. **Non-Compliances and Sanctions:** In case of non-compliances, the ICS shall take corrective or mitigating measures.
    - Procedures for implementation of sanctions will be defined in case of noncompliance.
    - Sanctions have to be documented (list of farmers issued sanctions, documentation of identified non-conformities in the files).
    - Farmers who have used prohibited inputs on their farms must undergo again the full conversion period (if they remain in the group). In such cases, it has to be checked whether the farmers have already delivered produce and whether this (now no longer certified) produce has been mingling with other produce, if this has been the case, the accredited Certification Body needs to be notified immediately and the mingled produce kept separate until further instructions.
16. **Training of ICS Personnel:** A competent person will train each internal inspector annually. The date of the training, list of participants and content of the training of all ICS staff needs to be documented.
17. **Training of Farmers:**The ICS manager will organize regular training to the Farmers in the group. Each farmer needs to receive at least one initial advisory visit by the extension service or in an organized training and the list of participants and content of the training needs to be documented.
18. **Buying Procedures:** To ensure genuineness of the products from the group. the following minimum requirements should be followed during buying:
    - The status of the farmer in the group should be checked.
    - The supplied amount should be compared with the harvested amount and estimated yield. In case of doubt, the produce is kept apart until clarified by the ICS manager.
    - The delivered quantity of the product will be registered in the purchase record.

- ☆ Farmer will be issued a receipt duly signed by the purchase officer stating the quantities of the product delivered with date.
- ☆ All documents have to indicate the status of the certified product (organic or in conversion).
- ☆ Bags should be labeled as 'organic' or as 'in-conversion'.

19. **Storage and Handling Procedures:** The purchase or the warehouse manager during the handling of produce shall check the document to ensure the compliance with the NPOP standards. The following are the minimum requirement that will be followed during storage and handling:
    - ☆ Identification of the product at all stages of product flow during transition.
    - ☆ Segregation of organic products from in conversion products.
    - ☆ Fumigation of containers, irradiation/ionization *etc.* are prohibited.
    - ☆ The location in the warehouse during storage must be labeled as 'organic' or 'in-conversion'
20. **Processing:** During the handling of the produce, the documentation must be checked for compliance with the NPOP standards.
    - ☆ The accredited Certification Body will inspect Central Processing Units.
    - ☆ Ingredients and processing aids must be used as defined in NPOP standards.
    - ☆ During the product flow (transition), the products should be separated from non-organic products.
    - ☆ The processing steps will be documented.

4.8 India Organic Trade Mark Logo

4.8.1 India Organic Logo

A trademark – "India Organic" is granted on the basis of compliance with the National Standards for Organic Production (NSOP). Only such exporters, manufacturers and processors whose products are duly certified by the accredited Certification Bodies, are eligible for grant of licence to use the logo.

As per the provisions of FTDR Act the products will be allowed to be exported as "Organic Product" only if it is produced, processed and packed under the Certification Trade Mark issued by Accredited Certification Bodies authorized by the NAB, constituted under the provisions of the NPOP. NAB is the sole, absolute, and exclusive owner of the Certification Trade Mark. The Accredited Certification Body, while granting certification to an Applicant is merely acting as an agent of the NAB.

4.82. Grant of License

If, after having regard to requisite skill, resources, production, processing previous performance and antecedents relevant to the issuance of the licence, the Accredited Certification Body is satisfied that the applicant is fit to use the Certification Trade Mark, the Accredited Certification Body issues a licence authorizing the use of the Certification Trade Mark in respect of the product or class of products manufactured by the applicant in respect of the process employed in any production, manufacture or work, subject to such terms and conditions as specified in these regulations for a period not exceeding one year.

The Applicant on receipt of the license shall be entitled to use the Certification Trade Mark and restrict its use to such products or services, which will meet the norms and standard specifications of the products, set out in the NPOP. The Certification Trade Mark may be affixed to the products and/or used on packaging or promotional material or in the context of advertising activities.

In the event of a withdrawal of the right to use the aforesaid Certification Trade Mark, the certificate or the Licence shall be returned to the Accredited Certification Body.

4.8.3 Withdrawal of License

Any licence granted by the Accredited Certification Body may be suspended or cancelled by it, if it is satisfied that:

i. The products marked with the Certification Trade Mark do not comply with the related norms and procedures under NPOP; or
ii. The licensee had used the Certification Trade Mark in respect of a process which does not comply with the NPOP; or
iii. The licensee failed to provide reasonable facilities to the Accredited Certification Body to enable them to discharge the duties imposed on them; or
iv. The licensee has failed to comply with any of the terms and conditions of the license.

4.8.4 Surveillance and Regular Review

The grant of a licence shall be followed by surveillance visits determined by the Accredited Certification Body. The surveillance visits may be without notice to the applicant to ensure that the systems and procedures already assessed are being maintained. The special reassessment visit shall be necessary where an applicant fails to observe the conditions of the licence or where there have been significant changes in the organization of the applicant. The licensee shall be liable for the costs of such special visits.

4.8.5 Responsibility of Licensee

Licensee on grant of a licence to use of Certification Trade Mark shall:

i. At all times comply with the requirements of the license as set out therein and comply with these Regulations or any amendments thereto;

ii. Only claim that it is holding a license in respect of the capability which is the subject of the license and which relates to the products or processes in accordance with the license requirements;

iii. Not use the license in any manner to which the Accredited Certification Body may object and shall not make any statement concerning the authority of the applicant's use of the license which in the opinion of the Accredited Certification Body may be misleading;

iv. Submit to the Accredited Certification Body for approval the form in which it proposes to use its license or proposes to make references to the license;

v. Upon suspension or termination of the license, however determined, discontinue its use forthwith and withdraw all promotional and advertising matter which contains any reference thereto;

4.9 Regulation of Term 'Organic'

Several major developments in the field of standards and regulations occurred in 2009. The new EU regulation on organic production and the Canadian organic standard came into effect. Moreover, the Australian domestic organic standard was also implemented. Canada and the U.S. concluded the world's first fully reciprocal agreement between regulated organic systems, and the EU established procedures for approving certification bodies from outside the EU. These developments are expected to ease trade in organic products and boost the future growth of the sector. The number of countries with regulation of organic standards has risen to 73, while 16 countries are in the process of drafting legislation.

4.9.1 United States (U.S.)

The US Department of Agriculture (USDA) organic products are strictly regulated. They should have strict production and labeling requirements and the organic products must meet the following requirements:

- ☆ Must be produced without prohibited methods such as genetic engineering, ionizing radiation, or sewage sludge.
- ☆ Must be produced using permitted substances as mentioned in the National List of Allowed and Prohibited Substances.
- ☆ Must be supervised by a certifying agent authorized by USDA NOP, following all USDA organic regulations.

4.9.2 India

The use of the 'Indian Organic' logo is regulated by Section 6 of the NPOP. This logo may only be used on products that have been duly certified by accredited inspection and certification agencies as satisfying all organic standards prescribed in the NPOP. Although the use of certifiers' mark is voluntary, it is mandatory to place the 'Indian Organic' logo on all certified organic products originating from India.

What Does Labelling of a Product as 'Organic' Mean?

In the U.S., when a food or beverage product is labelled as 'organic,' it means that the product has been grown, produced, examined, and certified to be following the organic standards as mandated by the NOP, a program of the USDA.

Only the USDA can authorize a company to market and label its food/beverage as organic. If a company is authorized to label a product as USDA Organic, it implies it has met the standards of USDA NOP, which include the following:

- ☆ **Pesticide-Free:** Foods are produced without the usage of most conventional pesticides.
- ☆ **Fertilizer-Free:** Foods are produced without the usage of fertilizers made with synthetic ingredients or sewage sludge.
- ☆ **Bioengineering and Radiation-Free:** Foods produced are devoid of bioengineering or ionizing radiation.
- ☆ **Antibiotics and Growth Hormone-Free:** Organic meat, poultry, eggs, and dairy products from animals must not contain antibiotics or growth hormones.
- ☆ **Sustainable Practices:** The production process must employ renewable resources and conserve soil and water to improve environmental quality for succeeding generations.

Distinct Organic Label Statements for different Levels of Organic

There are different organic label statements for different levels of organic foods as mentioned below:

- ☆ **'100 per cent Organic':** 100 per cent of ingredients are organic, with 100 per cent organic processing.
- ☆ **'Organic':** 95 per cent or more of ingredients are organic, some chemical additives approved by USDA may be used in processing.
- ☆ **'Made with Organic Ingredients':** Certain ingredients are organic. This label statement is not a USDA standard labeling statement and cannot be used outside the ingredients panel on the label. However, it can be made if a certifying agent approved by the USDA has verified the claim of some ingredients as organic.

It is necessary that the product must be labelled with the correct organic statement. Before starting the usage of the term 'organic' on a product label sold in the U.S., the USDA must give your business the official certification and approval.

4.10 Legal Requirements for Organic Labels

If a producer/seller wants to label their business products as 'organic', they need to be certified first and must fulfil the requisite standards that apply to organic products in their respective country and comply with consumer law. However, a business cannot label a product 'organic' if it does not comply with the strict criteria as set by the law in their respective countries. Businesses that label their products as 'organic' should be able to verify their claims, whenever questioned so by consumers.

The certification can be obtained with the completion of applications, audit procedure, and the process of getting approval for organic labelling. After certification, producers can legally label their products with an 'organic' symbol, logo or trademark to indicate such status. Labels such as '100 per cent organic,' 'made using organic ingredients' or 'certified organic' target those customers who are buying the product for its organic status as much as for the specific product. When products are labelled 'organic,' sellers know that consumers find it a very attractive proposition.

4.10.1 Australia

In Australia, the prescribed standard for organic products is determined by Australian Standard (AS 6000). This guideline was developed with a view to standardize practices within the organic industry. It aims to operate as a standard yardstick across Australia that regulates how people can grow, produce, distribute, market and label organic products. Through a uniform regulatory mechanism, consumers can comfortably distinguish whether a product is organic or not, without confusion. A seller of an organic product should ensure that his operations are compliant with this standard, while a buyer of organic products should watch out for the label that mentions whether the product complies with the AS 6000.

Certification is provided by several different private organizations in Australia that are accredited by the Department of Agriculture, Fisheries and Forestry. The consequences for falsely labelling products as 'organic' when they are not certified can be quite serious. The Australian Consumer Law (ACL) protects consumers in Australia from false or misleading representations made by sellers. Consumers have every right to know whether what they are buying is organically sourced and produced, be it from an ethical standpoint, or health or religious reasons.

4.10.2 United States (U.S.)

Certified organic foods are manufactured in accordance with the federal standards determined by the USDA NOP. These standards were followed in 2002 after the Organic Foods Production Act of 1990 and continue to be interpreted and developed by the National Organic Standards Board, a federal advisory committee appointed by the Secretary of Agriculture. Organic standards comprise many factors such as soil quality, animal raising, pest and weed control, and use of input materials.

Apart from determining requirements for growing, processing and handling organic agricultural products, the NOP also sets labelling requirements for such products. Labelling requirements are based on the percentage of organic ingredients in a product. Products labelled '100 per cent organic' should contain ingredients and processing aids – excluding water and salt – that are only organically produced. Any other ingredients or additives are not permitted.

Products labelled 'organic' should contain a minimum of 95 per cent organically produced ingredients (excluding water and salt). Any residual ingredients should comprise non-agricultural substances that are included on the NOP National List of Allowed and Prohibited Substances.

Products that fulfil either of these labelling requirements may display the necessary phrases, as well as the percentage of organic content, on the product's primary display panel. Organic products bearing such labels should be grown, handled and processed without the use of pesticides or other synthetic chemicals, irradiation, fertilizers made with synthetic ingredients or bioengineering. The USDA seal and the organic certifying agent seal/mark may be displayed on product packages and in advertisements.

To use the phrase 'made with organic ingredients' and mention up to three of the organic ingredients or food groups on the primary display panel, processed products should contain minimum 70 per cent organic ingredients. For instance, a soup produced with minimum 70 per cent organic ingredients and only organic vegetables may claim 'soup made with organic peas, potatoes, and carrots' or 'soup made with organic vegetables.' Processed products containing below 70 per cent organic ingredients cannot use the term 'organic' anywhere on the primary display panel. However, they can identify specific ingredients that are organically produced on the ingredients statement on the information panel.

Akin to other organic products, processed products labelled 'made with organic ingredients' cannot be produced using any processes disapproved by the NOP. The percentage of organic content and the certifying agent's mark may be used on the primary display panel. However, the USDA seal cannot be used anywhere on the package.

The NOP does not have any limitations with respect to the use of other truthful labelling claims such as 'no drugs or growth hormones used,' 'free range' or 'sustainably harvested.'

The label 'organic' can be used only by agricultural products that achieve organic certification. Such products include foods and beverages such as cheese, chocolate, cookies, juices, meats, milk, pasta, poultry, prepared sauces, soups, wines and alcoholic beverages, *etc.* When made of organically grown natural fibbers, fibre products such as clothing, bedding, and tablecloths can also be labelled 'organic'.

4.10.3 India

In India, the NPOP lays down the general principles, recommendations and specific requirements for the use of organic labelling and claims. The labelling should convey unambiguous and accurate information on the product's organic status. Following closely the IFOAM Basic Standards, four categories of products are identified for labelling purposes based on their organic composition:

- **Single-ingredient products with 100 per cent organic**: Raw or processed agricultural products containing 100 per cent certified organic ingredients (excluding water and salt, but including additives) may be labelled 'produce of organic agriculture' or such similar description.
- **Multi-ingredient products with minimum 95 per cent organic ingredients**: Raw or processed agricultural products containing a minimum of 95 per cent certified organic ingredients (by raw material weight, excluding water and salt, but including additives) may be labelled 'certified organic' or such similar description.
- **Multi-ingredient products with minimum 70 per cent organic ingredients:** Raw or processed agricultural products containing between 70 per cent and 95 per cent certified organic ingredients (by raw material weight, excluding water and salt, but including additives) may be labelled 'made with organic ingredients' or a description akin to it on the principal organic agriculture and the law display, provided that the proportion of organic ingredients is clearly stated. Such products cannot be simply labelled 'organic'.
- **Multi-ingredient products with below 70 per cent organic ingredients**: Raw and processed agricultural products containing below 70 per cent certified organic ingredients may only contain indications that an ingredient is organic on the ingredient list, but cannot be labelled 'organic'.

In all cases, the person or company legally responsible for the manufacture and processing of an organic product shall be identified and no such product can be labelled as GE (genetic engineering) or GM (genetic modification) free to prevent possible misleading claims about the end-product. Moreover, all raw materials of multi-ingredient products, including additives, should be listed on the product label as per their weight percentage.

The NPOP accreditation system has been recognized by the USDA as satisfactory to meet the accreditation requirements of the US NOP. This implies that inspection and certification agencies accredited by India's National Accreditation Body are qualified to certify Indian organic products as compliant with the standards of the US NOP. The USDA seal may be appended to products certified as such, which can then be exported to the US as organic.

4.11 Labelling of Various Categories of Organic Products under National Organic Standards (NOS)

An overview of labelling the various categories of organic products is as follows:

- **Primary Display Panel:** The part of the package most likely to be seen by buyers at the time of purchase.

- **Information Panel:** This includes that part of the label of a packaged product that is immediately adjacent to and to the right of the primary display panel as observed by a person facing the primary display panel, unless any other section of the label is designated as the information panel because of package size or its attributes.
- **Ingredients Statement Panel**: The list of ingredients contained in a product shown in their common and usual names in the descending order of predominance.
- **Other Panel**: Any panel other than the primary display panel, information panel, or ingredients statement panel.
- The certifying agent will review and approve each of the product labels to ensure compliance.

4.12 Produts Labelled as 'Organic,' but are't

The manufacturer/seller can face serious legal repercussions if products labelled 'organic' are, in fact, not organic. Therefore, businesses cannot simply make statements or representations that are incorrect or likely to mislead the consumer's mind.

The sellers should also be cautious not to copy another product's 'certified organic' logo without their permission. This not only constitutes an infringement of consumer law, but also leads to infringement of someone else's trademark. Businesses are liable for misleading or deceiving customers even if they didn't do it deliberately. It is imperative for manufacturers/sellers to ensure compliance with the requisite standards before labelling their products as organic. As a precaution, producers/sellers should keep detailed records of their production processes, farming techniques and ingredients used in case of future disputes.

To sum it up, if a company manufactures a product and wants to claim that the product or its ingredients are organic, the final product needs to be 'certified.' If the product is not certified, the company should not make any organic claim on the primary display panel or use the USDA organic seal anywhere on the package. However, there are some operations that are exempted from certification, including organic farmers whose revenues are $5,000 or less. Only on the information panel, the company can identify the certified organic ingredients as organic and the percentage of organic ingredients.

- **Accreditation:** Accreditation means a procedure adopted by the National Accreditation Body for ascertaining the competence of a Certification Body to certify organic farms, products and processes as per the National Standards for Organic Products.
- **Accreditation Body:** The Accreditation Body shall be the agency set up by the Steering Committee for National Programme for Organic Production for accrediting Certification Body.

Table 4.1. Total Area under Organic Certification Process during Last 6 Years (cultivated + Wild Harvest) (in ha)

Sl.No.	State Name	2015-16	2016-17	2017-18	2018-19	2019-20	2020-21
1	Madhya Pradesh	2275567.10	2292697.39	1156881.40	918303.08	1161015.03	1637730.46
2	Rajasthan	553447.70	539522.12	442133.72	632701.23	539245.81	481862.38
3	Maharashtra	266299.24	292391.78	304074.81	261571.74	293135.19	371798.28
4	Chhattisgarh	180924.94	179752.14	191464.66	206180.71	208392.80	286684.52
5	Himachal Pradesh	1358449.24	14376.72	170153.47	203847.50	204836.35	203736.47
6	Jammu and Kashmir	54515.01	181608.32	180870.34	187002.89	215275.95	192769.82
7	Karnataka	133647.27	81948.81	105515.02	104962.37	170418.49	174423.56
8	Uttar Pradesh	106292.39	101459.95	192734.40	205980.82	132031.67	159307.73
9	Gujarat	80421.40	70495.05	85400.71	94708.69	95207.58	147866.41
10	Odisha	109224.05	99736.17	117910.30	127851.77	115676.68	96306.88
11	Uttarakhand	99900.39	93586.42	104134.66	41409.55	43647.02	82210.20
12	Jharkhand	77048.73	36813.95	51187.93	58116.87	64254.18	81661.70
13	Sikkim	75851.21	75218.28	76076.18	75798.92	75717.65	75729.66
14	Kerala	44788.50	43701.88	34160.14	40911.24	47575.29	48364.18
15	Tamil Nadu	19529.79	10775.69	20070.51	26546.83	36766.59	41618.86
16	Meghalaya	4609.42	9629.60	40335.66	48409.74	45382.40	38376.39
17	Andhra Pradesh	93350.73	172783.03	184748.65	37409.72	42101.87	36801.36
18	Bihar	91.70	679.20	695.80	3519.51	22712.55	29902.54
19	West Bengal	17890.41	5176.03	5811.48	20989.65	6392.05	21002.61
20	Punjab	17577.20	17648.53	18000.77	25524.58	25637.95	18637.50
21	Assam	28493.24	23930.40	28071.81	28234.67	26753.67	18470.84
22	Goa	16957.59	15762.43	15698.98	20964.80	20786.66	18222.16
23	Nagaland	6186.93	4699.93	8839.86	8268.56	14254.97	14790.38

Sl.No.	State Name	2015-16	2016-17	2017-18	2018-19	2019-20	2020-21
24	Manipur	251.40	241.40	5397.90	7460.82	14990.07	14724.92
25	Arunachal Pradesh	72485.26	72311.27	6179.69	9246.94	10657.66	13114.12
26	Mizoram	213.80	210.00	998.95	7039.89	10029.89	13038.89
27	Telangana	10355.59	9687.84	8919.82	8759.52	8742.28	6865.56
28	Tripura	203.56	203.56	2251.19	2534.52	3539.18	6521.31
29	Haryana	4889.21	5031.76	6912.40	5998.58	6155.75	4903.06
30	Lakshadweep	895.52	895.52	895.51	895.51	895.51	895.51
31	Ladakh	0.00	0.00	0.00	0.00	64.22	817.85
32	Pondicherry	2.84	2.84	2.84	2.84	23.65	23.65
33	New Delhi	23.03	9.23	9.23	0.72	0.72	5.17
34	Andaman and Nicobar Islands	0.00	0.00	0.00	7484.00	7484.00	0.00
	Total	**5710384.00**	**4452987.00**	**3566538.00**	**3428638.77**	**3669801.33**	**4339184.94**

Source: https://apeda.gov.in/apedawebsite/organic/data.htm#Total_farm_area_2021

- ☆ **Accredited Certification Body:** An organization with legal entity complying with NPOP accreditation criteria and recognized by the National Accreditation Body for certifying organic products and for granting the right to use the Certification Trade Mark to the operators on behalf of the Accreditation Body.
- ☆ **Accredited Programme:** The accredited programme is the programme of the Certification Body that has been approved by the Accreditation Body on the basis that it is in compliance with the provisions of the National Programme for Organic Production.
- ☆ **Applicant Body:** Applicant body shall mean the organization seeking accreditation.
- ☆ **Buffer Zone:** A clear defined and identifiable area boarding an organic production/site from that of conventional production unit.
- ☆ **Certificate of Accreditation:** The Certificate of Accreditation is a document issued by APEDA, on behalf of the National Accreditation Body (NAB), to the Certification Body certifying that the accredited Certification Body is compliant with the standards as envisaged under the National Programme for Organic Production and is competent to certify producers as per the standards specified in the National Standards for Organic Production.

Chapter 5

Soil Improvement and Amendments

5.1 Soil Amendments

Soil amendments are organic or inorganic matter added to the soil to improve texture, water retention, drainage or aeration. Sandy or rocky soil requires amendments to improve the texture and add water retention properties. Clay soils require the addition of material to improve texture, aeration and drainage. Soil amendments come in a variety of sources.

Fertilizers improve the supply of nutrients in the soil, directly affecting plant growth. Soil amendments improve a soil's physical condition (*e.g.* soil structure, water infiltration), indirectly affecting plant growth.

Fresh organic matter can also be used, but extra nitrogen fertilizer will be needed to help it decompose. Fresh manure may contain too much uric acid, which can burn roots. It is best if manure is aged for at least six months. The best and easiest time to improve soil is to do an entire area at once, such as when planting a new lawn or landscape. The amendments should be spread evenly over the area and worked into the soil at least ten inches deep. A spade or spading fork works

A soil amendment is any material added to a soil to improve its physical properties, such as water retention, permeability, water infiltration, drainage, aeration and structure. The goal is to provide a better environment for roots. To do its work, an amendment must be thoroughly mixed into the soil.

best for small areas. A roto-tiller handles large areas. For existing beds, the soil can be improved every time something is planted.

For flower bulbs and root crops, the soil should be double dug to fifteen to eighteen inches deep. Double digging involves digging up the top eight inches of soil and piling it to the side of the hole. Soil should be worked when it is moist but not wet. Wet soil will be packed into hard clods, which are about like rocks. To check the soil for dampness, squeeze a handful of soil into a ball, then push your thumb into the ball. It should crumble, not dent like modeling clay. Dry soil is harder to work and the soil disintegrates into dust. Dust will turn into mud when watered, then dry like brick. Dry soil should be watered well then allowed to soak in for a day before tilling.

Soil should not be worked into dust. Leaving marble-sized to golf ball-sized clumps will give better aeration and drainage. Besides destroying the structure of soil, over working soil also kills off the mycorhizae which are beneficial fungi for moving nutrients from soil into roots. Roto-tilling may be required to incorporate amendments into the soil but more mycorhizae will survive if the soil is loosened with a spading fork. Simply stomp the tines into the soil and tilt back the handle to lift and fracture the soil.

5.2 Inorganic Soil Amendments

Inorganic amendments are **either mined or created by man**. Some examples include: dolomite lime, gypsum, sand, pea gravel, vermiculite and perlite. One reason to amend soil with inorganic amendments is to improve porosity and thus increase water and air permeability of the soil.

Table 5.1. Common Inorganic Amendments

Amendment	*Function/Purpose*
Sand or Profile Soil Conditioner:	to improve aeration and drainage
Lime or Sulfur:	to raise or lower pH
Vermiculite, a natural balsamic mineral and perlite made from heated amorphous volcanic glass	increases aeration, improves texture and assists in water retention.

5.3 Organic Amendments

Various types of composts and organic soil amendments play a pivotal role in sustaining the soil fertility and plant health for their intrinsic/acquired abilities to perform an array of functions such as soil humus formation, plant growth promotion, biological control of pest organisms, improvement of soil tilth and structure, *etc.* Varieties of carbon enriched compounds present in organic amendments influence the dynamics and activities of beneficial microbiota in soil. Usual addition of organic residues can increase soil physical fertility, mainly by improving aggregate stability, decreasing soil bulk density and increasing water holding capacity. Organic residues are particularly important for plant nutrition as they provide plants not only with available nitrogen, phosphorus, potassium and different secondary and micro-nutrients but also with physiologically active substances that stimulate plant growth. Soil suppressiveness towards soil-borne plant pathogens can be achieved by altering the diversity and balance of microorganisms in soil by repeated application of different types of compost and decaying organic matter. Application of different types of organic amendments in soil for long run will not only have an impact on sustainable economic development, but also contribute to a resilient agricultural ecosystem.

Organic amendments are those that are originated mostly from plant or animal source and most of them are macrobiotic, crude and bulky in nature. Organic amendments often contain biologically-active molecules such as vitamins, growth regulators, toxins and low molecular mass humic substances which can affect the soil microorganisms. The direct addition of organic amendments to crop soils can improve soil quality by affecting many parameters such as soil aeration, structure, drainage, moisture holding capacity, nutrient availability and microbial ecology Feasibility of using organic amendments such as compost, animal manures and organic industrial byproducts in order to suppress soil-borne plant pathogens has been well documented recently reviewed on a wide range of organic amendments used for sustainable cotton production system in Vertisol. However, the present review explicit the mode of action and role of organic soil amendments towards plant and soil health management in respect to low external input resilient agriculture.

5.4 Functions of Organic Materials

The organic materials most commonly used to improve soil fertility include farmyard manure (FYM), compost, crop residues, city refuse, rural compost, poultry manure, green manures, bio-gas spent slurry, microbial preparations, vermicompost, agro-industrial wastes (oilseed cakes, press mud cake, rice husk and biomass ashes), residues from processing of animal products (blood, horn-and bone-meal), sewage sludge *etc.*

Compared to the farm using no organic inputs, the regular use of organic amendments can result in:

- Increased SOC;
- Increased active C pool of micro-organisms;
- Improved soil WHC and nutrient holding capacity (NHC);
- Improved bulk density, structure, porosity and permeability, which can reduce runoff and erosion;
- Higher phosphorous buffering capacity;
- Less leaching of nitrate from the soil profile;
- Maintenance of soil pH to optimum levels;

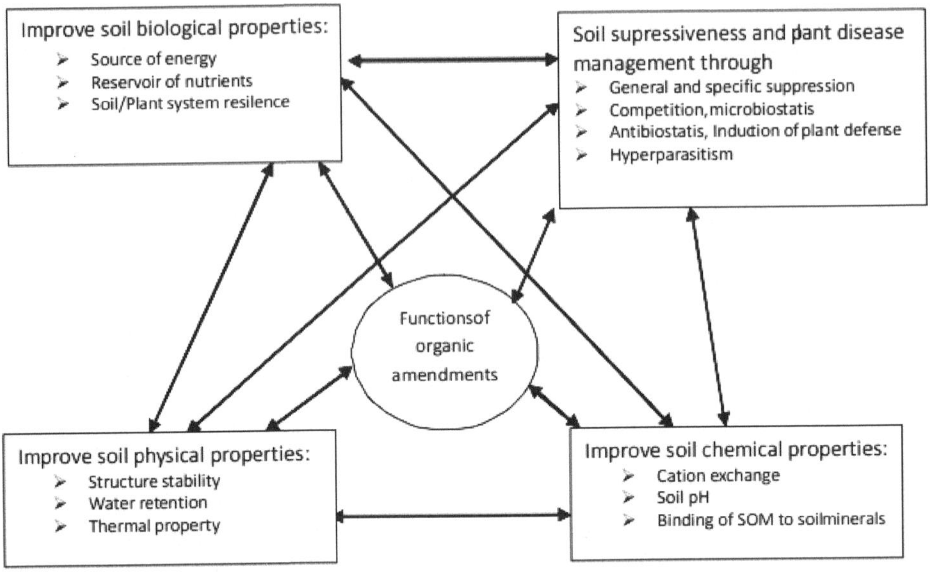

Figure 5.1. Functions of Organic Amendments.

☆ Stimulating soil biological components, affecting the quantity, diversity and functions of soil microbiota, thereby helping in soil aggregation, nutrient cycling, plant disease suppression *etc.*

Table 5.2. Common Organic Amendments

Amendment	*Function/Purpose*
Manures	to add nutrients
Compost	improves the texture of the soil, increases aeration, promotes both water retention and drainage as well as adds valuable nutrients necessary for plant growth.
Mulch	Mulch helps soil retain its moisture
Humus	Humus offers stability to imbalanced soil and rectifies deficiencies while maintaining its own balance
Peat Moss	It increases the ability of the soil to retain water and is ideal for sandy or rocky soil. It is ideal for stabilizing clay soils as well.
Wood products like woodchips and shavings	To improve soil, but may create a nitrogen deficiency.

5.5 Soil Drainage and Aeration

Drainage and aeration is improved by soil amendments, but there must be some place for the water to go. Yards should be graded so surface water drains off. Drainage sensitive plants such as Japanese maples, daphne and heather should be planted on a mound at least a foot tall or on a slope. Vegetables are best grown in raised beds, which provide better drainage and warmer soil in the spring for an earlier start.

In mostly flat yards, subsurface drainage may have to be provided. Perforated plastic drainpipe can be buried about a foot deep. It is important that drainpipes slope evenly so dirt doesn't clog up the low spots. Drain pipes can drain into down spouts if allowed by local building codes.

Aeration can be improved on existing lawns and beds without tearing up the soil. Sod core aerators cut out plugs of soil and leave holes about six inches apart and four inches deep. Root feeders can be used around trees and shrubs to create holes eighteen inches deep. These holes allow air and water to quickly penetrate deep into the soil. The holes will stay open much longer if sand or Profile Soil Conditioner is raked into the holes to fill them.

Chapter 6

Weed Management in Organic Farming

6.1 Introduction: An Overview

Weed, general term for any plant growing where it is not wanted. Ever since humans first attempted the cultivation of plants, they have had to fight the invasion by weeds into areas chosen for crops.

Some unwanted plants later were found to have virtues not originally suspected and so were removed from the category of weeds and taken under cultivation.

Weeds compete with crop plants for water, light, and nutrients. Weeds of rangelands and pastures may be unpalatable to animals, or even poisonous; they may cause injuries, as with lodging of foxtails (Alopecurus species) in horses' mouths; they may lower values of animal products, as in the cases of cockleburs (Xanthium species) in wool; they may add to the burden of animal care, as when horses graze in sticky tarweeds (Madia species). Many weeds are hosts of plant disease organisms. Examples are prickly lettuce (Lactuca scariola) and sow thistle (Sonchus species) that serve as hosts for downy mildew; wild mustards (Brassica species) that host clubroot of cabbage; and saltbrush (Atriplex species) and Russian thistle, in which curly top virus overwinters, to be carried to sugar beets by leafhoppers. Many weeds are hosts of insect pests, and a number are invasive species.

- Weeds are the plants, which grow where they are not wanted (Jethro Tull, 1731).
- In 1967 the Weed Science Society of America defined a weed as "a plant growing where it is not desired" (Buchholtz, 1967).
- In 1989 the Society's definition was changed to define a weed as "any plant that is objectionable or interferes with the activities or welfare of man welfare of man" (Humburg Humburg, 1989, p. 267;, 1989, p. 267; Vencill Vencill, 2002, p. 462)., 2002, p. 462).
- The European Weed Research Society defined a weed as "any plant or vegetation, excluding fungi, interfering with the objectives or requirements of people" (EWRS, 1986).
- The Oxford English Dictionary (Little et al., 1973) defines a weed as a "herbaceous plant not valued for use or beauty, growing wild and rank, and regarded as cumbering the ground or hindering the growth of superior vegetation."
- "A plant out of place or growing where it is not wanted" W.S. Blatchley 1912
- "A plant that is growing where it is desired that something else shall grow" A.E. Georgia 1916
- "The obnoxious plants are known as weeds" W.W. Robbins et al. 1942
- "Those plants with harmful or objectionable habits or characteristics which grow where they are not wanted, usually in places where it is desired that something else should grow" W.C. Muenscher 1946
- "Higher plants which are a nuisance" J.L. Harper 1960
- "A plant growing where we do not want it" E.J. Salisbury 1961
- "A plant growing where it is not desired; or a plant out of place" G.C. Klingman 1961
- "A plant whose virtues have not yet been discovered" R.W. Emerson 1912
- "Weeds have always been condemned without a fair trial" F.C. King 1951

Weeds interfere with a variety of human activities, and many methods have been developed to suppress or eliminate them. These methods vary with the nature of the weed itself, the means at hand for disposal, and the relation of the method to the environment. Usually for financial and ecological reasons, methods used on a golf course or a public park cannot be applied on range-land or in the forest. Herbicide chemicals sprayed on a roadside to eliminate unsightly weeds that constitute a fire or traffic hazard are not proper for use on crop land. Mulching, which is used to suppress weeds in a home garden, is not feasible on large farms. Weed control, in any event, has become a highly specialized activity. Universities and agricultural colleges teach courses in weed control, and industry provides the necessary technology. In agriculture, weed control is essential for maintaining high levels of crop production.

Farmers have struggled with the presence of weeds in their fields since the beginning of agriculture. Weeds can be considered a significant problem because they tend to decrease crop yields by increasing competition for water, sunlight, and nutrients while serving as host plants for pests and diseases. Since the invention of herbicides, farmers have used these chemicals to eradicate weeds from their fields. Using herbicides not only increased crop yields but also reduced the labor required to remove weeds. Today, some farmers have a renewed interest in organic methods of managing weeds since the widespread use of agro-chemicals has resulted in purported environmental and health problems. It has also been found that in some cases herbicide use can cause some weed species to dominate fields because the weeds develop resistance to herbicides. In addition, some herbicides are capable of destroying weeds that are harmless to crops, resulting in a potential decrease in biodiversity on farms. It is important to understand that under an organic system of weed control, weeds will never be eliminated but only managed. Farmers who wish to become organically certified are restricted from using synthetic herbicides for weed control.

6.2 Weeds, Enemy or Ally?

Weeds are often seen as the enemy that has to be controlled, preferably eliminated. Good farmers are often viewed as those that have nothing on their farms except crops. Weeds are seen as an indication of failure. Things in agricultural/ ecological systems are rarely that simple however. An example of the complexity of farm ecosystems comes from cereal crops in the South of England. Increased use of herbicides in the crop and non-crop areas, such as hedgerows, resulted in a dramatic reduction in the number and type of weeds. This in turn caused an even greater reduction in the number of species, and populations of, beneficial insects (those that kill crop pests). This resulted in an increase in pest populations, requiring more pesticides to be used, resulting in increased costs. The effect of using more herbicides was pest outbreaks! This kind of effect is neither intuitive nor easy to predict, but is common in agro-ecosystems. By not spraying field margins, and growing one meter wide strips of tussocky grasses such as Yorkshire fog and coltsfoot at 100 m intervals across the field, beneficial insects were reintroduced into the crop and kept the numbers of crop pests below economic damage levels, without the use of pesticides. Weeds are therefore both an enemy, and an ally. An organic system that aims for the total eradication of weeds is likely to run into difficulties, and vice versa, abandonment of weed management will lead to severe crop losses. A balance is therefore needed. Organic systems need to have a diverse range of plants in field margins, and while managing crop weeds, total elimination would be counter productive.

6.3 Scope of Biological Weed Control

Most weeds are introduced species, which occur at much greater population densities than they do in their natural habitat. Their pest status is often due to competition with crops and pastures, toxicity to livestock, alteration of the habitat in natural wilderness areas or in natural or artificial water bodies. In biological control, population densities of the weed are reduced by the introduction, from the country of origin of the weed, of specific natural enemies. The employment of biological control does not lead to eradication.

For biological control to be continuously successful small numbers of the host must always be present to attack those plants that survive and to prevent an increase· in their numbers. Biological control has most often been used against alien plants that have become weedy. Native plants are also increasingly becoming targets of biological control attempts. Insects constitute the largest group of natural enemies of weed plants. Mites have also been used occasionally in biological control.

Fishes, snails, ducks and mammals (manatee) have been used in the control of aquatic weeds. Weeds may also be damaged by a variety of pathogens (fungi, viruses, bacteria) or by parasitic plants (dodder, witch weed). Insects are important in biological control due to their great variety and numbers, good knowledge of the systematics, life history and plant associations and the availability of a range of natural enemies suited to particular ecological situations. Most insect species are intimately adapted to their host plants, with high degree of host specialization in certain taxa and cause conspicuous damage. Moreover they are easy to handle.

6.4 Organic Methods of Weed Management

6.4.1 Introduction

Organic weed management is a holistic system involving an entirely different approach to managing a farming system. The organic farmer is not interested in eliminating all weeds but wants to keep the weeds at a threshold that is both economical and manageable. A farmer who manages weeds organically must be intimately familiar with the type of weeds and their growth habits to determine which control methods to employ.

6.4.2 Soil Solarization

During summer and fall, organic farmers sometimes sterilize their soil through solarization. During this process, a clear plastic film is placed over an area after it has been tilled. Solarization works when the heat created under the plastic film, which is tightly sealed at the edges, becomes intense enough to kill weed seeds.

6.4.3 Mulch

Mulching or covering the soil surface can prevent weed seed germination by blocking light transmission preventing seed germination. Allelopathic chemicals in the mulch also can physically suppressing seedling emergence.

Organic mulches such as hay, straw, tree leaves, and wood shavings keep light-responsive weed seeds in the dark, physically hinder emergence of weed seedlings, and can provide shelter for ground beetles and other weed seed consumers. They also conserve soil moisture for crop production, maintain good soil tilth, prevent surface crusting, feed soil life, and sometimes provide slow-release nutrients. About three or four inches of hay or straw mulch can greatly reduce the emergence of broadleaf weed seedlings. Organic mulches are less effective against grassy weeds, and usually do not significantly hinder the emergence of perennial weeds from rootstocks, tubers, rhizomes, or bulbs.

One limitation for organic mulch is that manual application of mulch materials may be too labor-intensive for multiacre plantings. Bale choppers have been developed to mechanize application of hay or straw between wide-spaced crop rows or beds. Another approach to mulching at the farm scale has been the production of *in -situ* mulch in the form of high biomass cover crops. This entails no-till cover crop management and vegetable planting, which are most feasible where existing weed pressure is light to moderate, perennial weeds are scarce, and a transplanted or large-seeded vegetable will be grown in the cover crop mulch. Rollers, roll-crimpers, flail mowers, and undercutters are used to convert the mature cover crop into *in situ* mulch at the farm scale, leaving residues either chopped fine enough (flail mowing) or oriented parallel to future crops rows to permit mechanized planting. Gardeners can cut cover crops with a scythe, sickle, or weed whacker, and plant vegetables manually with a spade, trowel, or dibble.

Living mulches consisting of low-growing cover crops between cash crop rows are most workable for perennial fruit crops, especially tree fruits and grapes. Living mulches usually compete too strongly with vegetables, resulting in yield losses.

Black plastic film mulch effectively blocks emergence of most weeds, including perennials. They also eliminate the light stimulus for weed seed germination. However, these synthetic materials do not enhance soil quality, can interfere with infiltration of rainfall, and (unless a biodegradable material is used) require pickup and disposal at the end of the season. In addition, weeds often come up through planting holes, where they can be especially hard to control. Weed barriers (landscape fabric) last several seasons and can be especially useful for getting perennial crops established. Paper mulches used alone are less effective than plastic, but a paper mulch underneath hay or other organic mulch can enhance weed control over the organic mulch alone.

Clear plastic film mulch raises soil temperature much more than black plastic. During hot summer weather, it can effect soil solarization, another form of thermal weed control. Solarization kills emerging weeds, some soil-borne crop pathogens and insect pests, and even some weed seeds and vegetative propagules of perennial weeds. However, if conditions are too cool or cloudy to support effective solarization, the clear plastic can simply accelerate weed growth under the plastic layer by creating near-optimum temperatures.

6.4.3.1 Forms of Mulches

There are many forms of mulches available. Three common are described as under:

- ***Living mulch:*** A living mulch is usually a plant species that grows densely and low to the ground, such as clover. Living mulches can be planted before or after a crop is established. It is important to kill, till in, or otherwise manage the living mulch so that it does not compete with the actual crop.
- ***Organic mulches:*** Such materials as straw, bark, and composted material can provide effective weed control. Producing the material on the farm is recommended since the cost of purchased mulches can be prohibitive, depending on the amount needed to suppress weed emergence. An effective but labor-intensive system uses newspaper and straw. Two layers of newspaper are placed on the ground, followed by a layer of hay. It is important to make sure the hay does not contain any weed seeds.
- ***Inorganic mulches:*** Materials such as black polyethylene have been used for weed control in a range of crops in organic production systems.

6.5 Mechanical Weed Management

Managing weeds mechanically is both time-consuming and labor-intensive but it is also one of the most effective methods for managing weeds. The choice of implementation, timing, and frequency will depend on the structure and form of the crop and the type and number of weeds. Cultivation involves killing emerging weeds or burying freshly shed weed seeds below the depth from which they will germinate. It is important to remember that any ecological approach to weed management begins and ends in the soil seed bank. The soil seed bank is the reserve of weed seeds present in the soil. Observing the composition of the seed bank can help a farmer make practical weed management decisions.

6.5.1 Stale Seedbed

The stale or false seedbed technique of flushing out weed seeds from the soil works by depleting the seed-bank. After the soil is cultivated two to three weeks before sowing, emerging weeds are killed by flaming or light cultivation. By helping to reduce the seed-bank. This technique reduces subsequent emergences of weeds.

6.5.2 Crop Rotation

Crop rotation has been at the heart of the organic weed management system since medieval times and has persisted well into the 20th century due to its proven effects on weed populations. The goal of a crop rotation is to create an unstable environment that discourages weeds from becoming established in the field. Deciding on the sequence of crops, a farmer must take into account the type of soil he or she is working with, the climate, and the crop.

Crop establishment and competition Make sure crops emerge first to give them a head start in their competition with weeds. Transplanting helps increase a crop's competitive ability since the plants are larger and easier to establish. Sow crops close together by reducing the row spacing. Since the crop will take up more space, it shades the weeds, reducing the weeds' ability to compete. Another technique involves increasing the seeding rate of a crop. This increases the competitive ability of the crop by increasing the odds that the crop will survive in greater numbers than the weeds.

Limiting the spread and introduction of weeds It is important to maintain proper sanitation on the farm to reduce the introduction and spread of weed seeds. There are several ways to keep weeds and weed seeds from entering the farm. First, any animal manure that will be used on the farm should be composted because weed seeds can pass through an animal's digestive system unharmed, it is important to compost the manure. Composting results in temperatures that become high enough to kill many weed seeds. Second, purchase certified seed that is guaranteed to be free of weed seeds. If you are a farmer interested in saving your own seed, be diligent about collecting clean seed so you do not contaminate your collections. Also make sure to remove weeds before they set seed. Once a weed is allowed to set seed, the number of weed seeds in the seed bank is increased. Last, keep tillage and other equipment clean when moving between fields to reduce the spread of weed seeds.

Allelopathy

Allelopathy is an alternative and organic approach to weed control that uses chemicals that are excreted from a plant to cause either direct or indirect harm to weeds by negatively affecting their germination, growth, or development. Nearby weeds can be affected by allelopathic chemicals entering the rhizosphere from the roots or the aerial parts of the crop plant. Crop residues from cover crops, such as fall rye, or other organic mulches can also be used to suppress weeds through such allelopathic interactions.

6.6 Biological Weed Control

Biological weed control involves using living organisms, such as insects, nematodes, bacteria, or fungi, to reduce weed populations. In nature, plants are

Biological control utilizes natural living organism, such as insects, herbivorous fish, other animals, disease organisms and competitive plants to limit their growth. In biological control method, it is not possible to eradicate weeds but weed population can be reduced. This method is not useful to control all types of weeds. Introduced weeds are best targets for biological control. Biological control is least harmful to the environment, have no residual effect, relatively cheaper, having comparatively long lasting effect, harmless to non-targeted plants, very effective in control of weeds in non cropped areas. Besides this some of the fish, snails and other animals convert weed vegetation into seafood.

controlled biologically by naturally occurring organisms. Plants become pests - and are labelled "weeds" - when they run rampant because their natural enemies become ineffective or are non-existent. The natural cycle may be interrupted when a plant is introduced into a new environment, or when humans disrupt the ecological system.

Biological control means, Utilization of natural living organism, such as insects, herbivorous fish, other animals, disease organisms and competitive plants to limit their growth. In biological control method, it is not possible to eradicate weeds but weed population can be reduced. This method is not useful to control all types of weeds. Introduced weeds are best targets for biological control.

6.6.1 Merits and Demerits of Biological Weed Control

6.6.1.1 Merits of Biological Weed Control

- ☆ Least harm to the environment
- ☆ No residual effect
- ☆ Relatively cheaper and comparatively long lasting effect
- ☆ Will not affect non-targeted plants and safer in usage
- ☆ It is very effective in control of weeds in non cropped areas
- ☆ Besides this some of the fish, snails and other animals convert weed vegetation into seafood.

6.6.1.2 Demerits of Biological Weed Control

- ☆ Multiplication is costlier
- ☆ Control is very slow
- ☆ Success of control is very limited
- ☆ Very few host specific bio-agents are available at present

6.6.2 Working Mechanism

Roots provide plants with water and nutrients. Some biological control agents attach to roots and thereby stunt plant growth. Some bacteria live on root surfaces and release toxins that stunt root growth. Many fungi infect roots and disrupt the water transport system, which reduces leaf growth. Beneficial insects and nematodes feed directly on the weed roots causing injury which allows bacteria and fungi to penetrate.

Plant leaves capture energy from the sun and store it as sugar. Insects that feed on leaves reduce the leaf surface available for energy capture. Fungi and bacteria that infect leaves reduce the ability of the leaf to make sugars. In either case, there is less energy available for weed growth. Whether through damage on roots or leaves, severe infestations of biological control agents can actually kill weeds, reducing their adverse effects on desirable plants.

Many weed species survive from year to year by producing seeds. Fungi or insects that attack seeds can reduce the number of weed seeds stored in the soil, which in turn can reduce the size of future weed populations. This lowers the effort needed to control the remaining emerging weeds.

Some bacteria and fungi applied as biological control agents do not survive from year to year. These organisms must be applied on an annual basis. This technique is called the "bioherbicide" strategy. With this tactic, biological agents are used a in manner similar to chemical herbicides.

Weeds introduced from foreign countries often require a different strategy. Insects and pathogens are collected in the area of origin and evaluated for release in North America. Insect agents often require a number of years to become fully effective. Their growth is often hindered by adverse climatic conditions. Long-term monitoring is needed to determine their effectiveness. The release of biological control organisms in this manner is termed the "classical" approach to biological control. Fungi that naturally spread and infect weeds can also be used in a classical biological control strategy. The cost of developing and conducting a biological control program varies with the target weed and the strategy selected.

Biological control of weeds will not eliminate the need to use chemical herbicides. Both of these tools need to be integrated with cultural practices, such as tillage and crop rotation, in the battle against weeds. By using Integrated Weed Management, the development of weeds that are resistant to biological or chemical agents can be slowed.

6.6.3 Biocontrol Techniques

Two main biocontrol techniques are used:

6.6.3.1 Classical Biocontrol

This is the most commonly used technique and involves the introduction of natural enemies from their native range into an exotic range where their host plant has become a weed. This approach creates a natural, self-sustaining balance between the weed and its control agent, similar to that found in the weed's native range.

Main objective of classical biological weed control is restoring balance between target alien weed and its natural enemies in the ecosystem by introduction of suitable, exotic bio-agent. Successful bio-agent reduce the weed population first then the Bio-agent population due to starvation of food. After some time the bio-agent population may recover. This process continues in cyclic fashion till the bio-agent and weed population gets established at a low level. This method is a slow operating and currently used in non cropped areas. In crop fields, the bio-agent will not get opportunity to work on host weed due to frequent use of insecticides and fungicides in modern agriculture. Otherwise *Cyperus rotundus* can be controlled in crop fields with moth *"Bactra verutana"* and selective bio-control of *Ludwigia parviflora* (water purslane) by *Haltica cyanea* (steel blue beetle) in rice fields.

Characteristics of Successful Bio-Agent

- **Host-specific:** Bio-agents should be host specific and they should not attack other economic plant spp. They should pass starvation test *i.e.* they prefer to starve to death rather feed upon other than host weeds. Lantana was controlled by *Teleonemia scrupulosa* insect bio-agent. But in India it is likely to damage teak (*Tectona grandis*) and sesame (*Sesamum indicum*). *Zygogramma bicolarata* is an effective leaf eating bio-agent against Parthenium (carrot grass). But it is found to attack sunflower in India.
- **Bio-agent hardiness:** Bio-agent should free from its own parasites and predators. Bio-agent should withstand starvation for short or long periods of food shortage when the target weed population is brought to low level. But carp can't survive even a short period of starvation.
- **Feeding habit:** Bio-agents are more efficient in controlling weeds if they attack either flowers or seeds of the weed or bore into the stems than root and leaf feeders. But root-feeding insects are more effective in controlling perennial weeds.
- **Ease of multiplication:** Bio-agent should have high rate and ease of natural reproduction. It is very important for insects, pathogens, snails and competitive plants. But it is not desirable with carp as its increased population compete with natural fish.

Kinds of Classical Bio-Agents

Bio agent may be either specific or non specific. Specific bio agent attack only one or two specific weeds, while non specific bio agent feed upon a variety of vegetation. Specific bio agents are insects, plant pathogens and competitive plants. Non specific bio agents are Carp fish, snails, and mites.

Six kinds of Bio-agents were used to control weeds

- **Insects:** These are largely host specific *i.e.* one insect spp is employed to destroy the only one weed sp. First successful example reported from Hawai in 1902 *Lantana camara* controlled by Moth, *Crocidosema lantana*. Insects that were found effective belong to Lepidoptera, Hemiptera, Coleoptera, Diptera.
- **Carp fish:** Certain fresh water Carp fish consume large quantities of aquatic weeds. Whiteamur (Chines grass carp) "*Ctenopharyngo donidella*" is promising spp for aquatic weed control. This can grow more than its body weight i.e. 5kg/year and attaining up to 50kg at its full size. Herbivorous fish are not food specific. Whereas the common carp (Cyprinus carpio) a non herbivorous fish used to control submerged aquatic weeds.

- ☆ **Plant pathogen:** Many fungi attack specific weed species. For instance *Acacia glauca* controlled by spore suspension of *Cephalosporium zonatum*. Skeleton weed (*Chondrilla juncia*) controlled by rust causing fungi, Puccinia chondrillana.
- ☆ **Competitive plants:** Certain plants sp are very competitive in suppressing specific weeds. Slender spike rush (*Eleocharisa cicularis*) aquatic plant can cover the canal bottom and it is not allowing to establish destructive tall weeds. Typha species can be controlled by *Panicum purpurascens* or *Brachiari amutica* (Para grass). Marigold has potential to displacing *Parthenium* spp. *Cassia sericea* also suppressed.
- ☆ **Slender spikerush:** *Brachiari amutica*, Marigold displacing Parthenium.
- ☆ **Snails:** The large tropical fresh water snail, *Marisa cornuarietis* feed on aquatic weeds. Marisa feed on roots of water hyacinth, water lettuce and leaves of Salvinia. *Marisa cornuarietis* feed on aquatic weeds.

Outstanding Example of Classical Biocontrol

- ☆ ***Lantana camera*:** Lantana was the first weed controlled successfully with certain insect bioagents in Hawaii. Of these *Crocidosema lantana*, a moth was found to be promising in destroying flowers and seeds of lantana. In Australia, three successful insect biocontrol agents are hispine beetles (*Octotoma scabripennis* and *Uroplata girardi*) and tingid or lantana bug (Teleonemia scrupulosa).
- ☆ **Prickly pear (*Opuntia* sp):** In Austrlia biocontrol of *Opuntia inermis* with a moth Cactoblastis cactorum. In Tamilnadu and Maharashtra 40,000 ha land was recovered from the weed *Opuntia delini* by releasing *Dactyloplius tomentosus*, a Cochineal scale insect.
- ☆ **Water hyclinth (*Eichornea crassipes*):** it is worldwide aquatic weed infested transplanted paddy fields including India. Hyacinth moth, *Sameodes albiguttalis* feed up on young leaves and apical buds. Besides this beetles *Neochetina bruchii* and *N. eichorniae* are also damaging the water hyacinth.
- ☆ **Salvinia (*Salvinia molesta*):** In Kerala (India) curculionid beetle *Cytrobagous salviniae* used to clear the fresh water courses and paddy fields. They feed on terminal buds and rhizomes and petioles of salvinia.

6.6.3.2 Non-classical biocontrol

This includes Innundative (releasing large numbers of the agent to control the weed *e.g.* mycoherbicides) and Augmentative approaches (mass rearing and release of large numbers of a control agent that cannot be utilised as a mycoherbicide).

Chapter-7

Organic Manure

7.1 Organic Manure: An Overview

Organic manures are essentially derived from either plant or animal by-products. The plant by-products are mainly cotton seed meal, fruit pomaces, leaf compost soybean meal, wood ash *etc* while the animal by-products includes the likes of blood meal, feather meal, fish meal and fish emulsion, leather meal *etc.* Many of these products are far too expensive to justify their use in other than very specialized horticultural applications. However, certain agro-wastes obtained from shade trees provide ample scope for use as organic amendment. These wastes have good nutritive value and can be used as an input during composting.

One of the most common types of pre-packaged alternative soil amendments is the manure- or compost-based blended fertilizer. Several of these products have national distribution, and many more enjoy a loyal regional following. Such products are typically analyzed at 2 to 5 per cent for each nutrient. Dried compost is used as a bulking agent, source of nutrients, and organic matter. It is blended with several of the materials including rock minerals and plant and animal by-products. Nearly all products of this class sell for prices about three times greater than their conventional fertilizer value, but may be quite effective in farm situations. However, farmers with access to other sources of manure or compost can realize substantial savings by relying on local manure resources.

An organic fertilizer replenishes the soil naturally, without leaving chemical residues and gives good crop yield under sustainable agricultural practices. The plants end up healthier and safer than chemical fertilizer. Sustainable agriculture would ideally produce good crop yields with minimal impact on ecological factors such as soil fertility under organic farming system. A fertile soil provides essential

Table 7.1. Nutritive Value of different Organic Compost/Manures

Organic	*N*	*P*	*K*	*Ca*	*Mg*	*S*	*Fe*	*Mn*	*Zn*	*Cu*	*Mo*
	(g kg^{-1})						*(mg kg^{-1})*				
Neem cake	30	1.0	15	10	3.2	2.3	2085	65	17	10	2.5
Cotton cake	35	2.9	13	4.0	3.9	1.5	169	14	37	9	1.6
Mustard cake	22	2.1	12	6.0	2.1	1.5	1062	27	21	18	2.0
Groundnut cake	68	3.6	12	2.0	3.3	2.1	667	37	48	16	5.4
Gingelly cake	60	3.9	11	10	5.0	1.9	1262	50	64	25	11.0
Castor cake	58	11	9.2	3.6	2.4	-	197	22	36	13	-
Farmyard manure	5	2.5	3.0	11	5.5	1.8	2745	466	59	29	11.0
Goat manure	31	1.8	13	18	7.0	1.5	1874	711	120	67	12.0
Poultry manure	30	2.3	17	23	6.0	3.1	2411	234	194	40	10.0
Leaf compost	10	1.0	16	21	3.1	2.1	656	177	12	11	6.0
Vermi compost	20	8.0	12	33	11	8.6	6130	221	44	36	1.6
Coir compost	12	0.6	12	5.0	4.2	1.0	950	50	16	6.0	1.0

Source: S. Hamza and V. Srinivasan, Planter's chronicle, 99, 4 (2003).

nutrients for crop plant growth, supports a diverse and active biotic community, exhibits a typical soil structure, and allows for an undisturbed decomposition. Crop productivity in organic farming is achieved by a combination of management strategies, among which the development of an efficient and effective nutrient cycle plays an important role. Organic farming seems to be more appropriate as it considered the important aspects like sustainable. Agriculture is the most important sector for ensuring food security, alleviating poverty and conserving the vital natural resources that the world's present and future generation will be entirely dependent upon for their survival and well being, in the name of development, the environmental resources have been beyond comprehension. Intensive use of inorganic fertilizers and pesticides has been an important tool in the drive for increased crop production. In fact more fertilizers consumption is a good indication of agricultural productivity but depletion of soil fertility is commonly observed in soils. Organic Farming seems to be more appropriate as it considered the important aspects like sustainable natural resources and environment. It is a production system, which favors maximum use of organic materials like crop residues, FYM, compost, green manure, oil cakes, bio-fertilizers, bio-gas slurry *etc.* to improve soil health from the different experiment, microbial fertilizers like Rhizobium, Azotobacter, Blue green algae, Azolla *etc.* have increased the yield and also played important role for minimizing the harmful effect of pesticides and herbicides. Organic farming is a practical proposition for sustainable agriculture if adequate attention is paid to this issue. There is urgent need to involve more and more scientist to identify the thrust area of research for the development of eco-friendly production technology.

7.2 Major Organic Sources and Transformations

Carbon present in soil is in the form of organic matter. The organic materials most commonly used to improve soil conditions and fertility include farm yard manure (FYM), animal wastes, crop residues, urban organic wastes (either as such or composted), green manures, bio-gas spent slurry, microbial preparations, vermicompost and biodynamic preparations. Sewage sludge and some of the industrial wastes also find application in agriculture.

For all organic matter, atmospheric carbon dioxide serves as the main source of carbon. Carbon dioxide is converted to organic carbon largely by the action of photoautotrophic organisms; the higher green plants on land and algae in aquatic habitats. Carbon is being contentiously fixed into organic form through the process of photosynthesis and once bound; the carbon becomes unavailable for use in the generation of new plant life. Carbon fixation involves a reduction of carbon dioxide by hydrogen donor NADPH (reduced form of the co-enzyme nicotinamide adenine dinucleotide phosphate, NADP) and the synthesis of carbohydrate from reduced carbon through complex cyclic mechanism called the Calvin cycle. Carbon dioxide constitutes only 0.03 per cent by volume of the earth atmosphere. It has been estimated that the vegetation of the earth's surface consumes some 90 billion kg carbon dioxide per annum, about one twenty - fifth of the total supply of the atmosphere and that the total supply of carbon dioxide would be completely exhausted in twenty years at the present rate of photosynthesis, if not replenished by decomposition of organic materials. As the availability of carbon dioxide on the earth's surface is very limited, it must be recycled. Upon the death of the plants and animals, microbiological metabolism assumes the dominant role in cyclic sequence. The dead tissues added to soil undergo decay and are transformed into microbial cells and a vast heterogeneous body of carbonaceous compounds. According to the different stages of decomposition, the soil organic matter becomes available in distinct fractions. Farm yard manure made from cattle dung, excreta of other animals, animal tissues and excretory products, and compost from rural and urban wastes, crop residues and green-manure are collectively designated as bulky organic manures because of their low contents of major nutrients, while materials like oil cakes, fish meal, animal meal, poultry manures, slaughter house wastes containing comparatively higher contents of plant nutrients are grouped under concentrated organic manures. In general organic manures containing up-to two per cent nitrogen are included in bulky category and those with more than two per cent nitrogen are treated as concentrated. Irrespective of source and composition, organic matter when added into the soil undergoes microbial decay and becomes the food for micro flora and fauna. Even the microbial cells serve as a source of carbon for succeeding generations of microscopic populations. A great variety of microorganisms live in soil which include bacteria, actinomycetes, fungi, algae and protozoa. In general the number per gram of soil is bacteria > actinomycetes > fungi > algae > protozoa. The average nutrient content of bulky manures given in Table 7.2.

Table 7.2. Average Nutrient Content of Bulky Manure

Manure	*Percentage Content*		
	N	P_2O_5	K_2O
Animal refuse	0.3-0.4	0.1-0.2	0.1-0.3
Cattle dung, fresh	0.3-0.5	0.3-0.4	0.3-0.4
Horse dung ,fresh	0.5 -0.5	0.4-0.6	0.3-1.0
Poultry manure, fresh	1.0-1.8	1.4-1.8	0.8-0.9
Sewage sludge, dry	2.0-3.5	1.0-5.0	0.2-0.5
Sewage sludge, Activate dry	4.0-7.0	2.1-4.2	0.5-0.7
Cattle urine	0.9-1.2	Trace	0.5-1.0
Horse urine	1.2-1.5	Trace	1.3-1.5
Human urine	0.6-1.0	0.1-0.2	0.2-0.3
Sheep urine	15-1.7	Trace	1.8-2.0
Ash, coal	0.73	0.45	0.53
Ash, household	0.5-1.9	1.6-4.2	2.3-12.0
Ash Wood	0.1-0.2	0.8-5.9	1.5-36.0
Rural compost, dry	0.5-1.0	0.4-0.8	0.8-1.2
Urban compost, dry	0.7-2.0	0.9-3.0	1.0-2.0
Farmyard manure, dry	0.4-1.5	0.3-0.9	0.3-1.9
Filter-press cake	1.0-1.5	4.0-5.0	2.0-7.0
Rice hulls	0.3-0.5	0.2-0.5	0.3-0.5
Groundnut husks	1.6-1.8	0.3-0.5	1.1-1.7
Banana, dry	0.61	0.12	1.00
Cotton	0.44	0.10	0.66

Source: Organic Farming for Sustainable Agriculture by A.K. Dhama, 1996, Agro Botanical Publishers (India).

7.3 Chemical Nature of Organic Matter and its Decomposition

Organic matter is a complex mixture of many chemical substances. These materials serve as precursors of soil humus.

The most commonly found and well understood organic compounds finding their way into the soil and are subjected to normal decomposition processes are:

- Carbohydrates (sugars, starches, hemicelluloses, cellulose, pectins, gums, mucilages, *etc.*;),
- Proteins, aminoacids, amins, *etc.*;
- Fats oil, waxes, resins, *etc*;

- Alcohol, aldehydes, Ketones, *etc*;
- Organic acid;
- Lignin;
- Compounds having ring structures (phenols, tannins, hydrocarbons),
- Alkaloids and compounds with organic bases — pyridine and purine, and 9 miscellaneous substances — antibiotics, auxins, vitamins, enzymes and pigments.

The inorganic constituents, usually present in ash containing major and minor elements amounting 12-15 per cent by weight.The chemical composition of organic matter given in Table 7.3.

Table 7.3. Chemical Composition of Organic Matter

Fractions	*Per cent (Dry weight basis)*
Sugars, starches, amino acids and urea.	2-30
Ammonium salts (hot/cold water solubles)	
Fats, oils, waxes. (ether/alcohol solubles)	1-15
Proteins	5-40
Hemicelluloses	10-3
Cellulose	15-6
Lignin	5-30
Mineral matter (ash)	5-30

Source: FAO Soil Bulletin (1987).

Organic matter while entering into the soil, undergoes decay and finally gets converted into carbon dioxide, water, nitrogen, sulphur oxide and keeps replenishing naturally the carbon, nitrogen and sulphur cycles. On ageing, the water soluble constituents of the plant tissues like proteins and minerals decrease and the percentage abundance of cellulose, hemicellulose and lignin rises. Compounds with linear chain structure are easily decomposed compared to those with cyclic (or) ring structures. Lignin is more resistant to microbial attack. The order of easy decomposition is protein > cellulose > fats, waxes > lignin.

Nearly all biological transformations that occur in soil are mediated by microbial activity. This means all the factors that influence microorganisms are important in the transformation of dead plant materials. The most important factors are temperature, soil moisture, rate of gas exchange, easy availability of nutrients and structure of the colloidal minerals materials. The age of the plant residues, it's lignin content, and the degree of disintegration of substrata also govern the rate of decomposition.

Microorganisms decompose the organic materials to obtain energy for growth and carbon for the synthesis of new cell material Carbon dioxide, methane, organic acids, alcohol and other oxidized and partly oxidized form of carbon may be metabolic wastes for one group of microorganisms. Whereas they may serve as energy and carbon source for other group. To decompose complex compounds, microorganism are required to liberate more number of enzymes. Single organic compound like cytoplasm of blue green algae are decomposed readily in two and half days by 50 per cent while corn stalk required two and half months.

Generally, if fresh organic matter with a carbon: nitrogen ratio (proportion of percentage of carbon to that of nitrogen) of greater than 30 is added to a soil, there is immobilization of nitrogen during the initial decomposition process. For ratios between 20 and 30, the mineralization and immobilization may be equal. Release of mineral nitrogen early in the process is caused when organic matter with a carbon: nitrogen ratio of less than 20 decomposes. The carbon: nitrogen ratio of stabilised humus is considered to be 10: 1. This critical ratio is a reflection of the dynamic equilibrium that results from the dominating microbial population.

The microbial protoplasm contains approximately carbon: nitrogen ratios of 5: 1, 10: 1 and 15: 1 for bacteria, fungi and actinomycetes respectively. Immobilization is a temporary locking up of the available plant nutrients.

Under both natural and agricultural conditions, an equilibrium is established between the humus being synthesized from fresh additions of organic materials and the native humus already present in soil. Under normal conditions the mineralization of native humus is compensated by synthesis of new humus.

Table 7.4. List of Organic Manures

Manure	*Percentage Content*		
	Nitrogen (N)	*Phosphoric Acid (P_2O_5)*	*Potash (K_2O)*
Coir pith	1.20	1.20	1.20
Blood meal	10-12	1.2	1.0
Press mud	1-1.5	4-5	2-7
Bone meal			
1) Raw bone meal	3-4	20-25	-
2) Steamed bone meal	1–2	25-30	-
Fish meal	4-10	3.9	0.3-1.5
Animal refuse	0.3-0.4	0.1-0.2	0.1-0.3
Cattle dung, fresh	0.4-0.5	0.3-0.4	0.3-0.4
Horse dung, fresh	0.5 -0.5	0.4-0.6	0.3-1.0
Poultry manure, fresh	1.0-1.8	1.4-1.8	0.8-0.9
Sewage sludge, dry	2.0-3.5	1.0-5.0	0.2-0.5
Sewage sludge, activate dry	4.0-7.0	2.1-4.2	0.5-0.7
Cattle urine	0.9-1.2	Trace	0.5-1.0
Horse urine	1.2-1.5	Trace	1.3-1.5

Manure	*Percentage Content*		
	Nitrogen (N)	*Phosphoric Acid (P_2O_5)*	*Potash (K_2O)*
Human urine	0.6-1.0	0.1-0.2	0.2-0.3
Sheep urine	1.5-1.7	Trace	1.8-2.0
Ash, coal	0.73	0.45	0.53
Ash, household	0.5-1.9	1.6-4.2	2.3-12.0
Ash, wood	0.1-0.2	0.8-5.9	1.5-36.0
Rural compost, dry	0.5-1.0	0.4-0.8	0.8-1.2
Urban compost, dry	0.7-2.0	0.9-3.0	1.0-2.0
Farmyard manure, dry	0.4-1.5	0.3-0.9	0.3-1.9
Filter-press cake	1.0-1.5	4.0-5.0	2.0-7.0
Rice hulls	0.3-0.5	0.2-0.5	0.3-0.5
Groundnut husks	1.6-1.8	0.3-0.5	1.1-1.7
Banana, dry	0.61	0.12	1.00
Cotton	0.44	0.10	0.66
Maize	0.42	1.57	1.65
Paddy	0.36	0.08	0.71
Tobacco	1.12	0.84	0.80
Pigeon pea	1.10	0.58	1.28
Wheat	0.53	0.10	1.10
Sugarcane trash	0.35	0.10	0.60
Tobacco dust	1.10	0.31	0.93
Tree leaves, dry			
Calotropis gigantea	0.35	0.12	0.36
Careya arborea	1.67	0.40	2.20
Cassia ariculata	0.98	0.12	0.67
Dillenia pentagyna	1.34	0.50	3.20
Madhuca indica	1.66	0.50	2.00
Pongamia pinnata	3.69	2.41	2.42
Pterocarpus marsupium	1.97	0.40	2.90
Terimalia chebula	1.46	0.35	1.35
Terminalia paniculata	1.70	0.40	1.60
Terminalia tomentosa	1.39	0.40	1.80
Xylia dolabriformis	1.37	0.30	1.61
Green manures, fresh			
Cowpea (*Vigna unguiculata*)	0.71	0.15	0.58
Sesbania aculeata	0.62	-	-

Manure	Percentage Content		
	Nitrogen (N)	Phosphoric Acid (P_2O_5)	Potash (K_2O)
Cluster-bea (*cyamopsis tetragonoloba*)	0.34	-	-
Horse-gram (*Dolichos biflorus*)	0.33	-	-
Mothbean	0.80	-	-
Green gram (*vigna radiate*)	0.72	0.18	0.53
Sunnhemp (*Crotalaria juncea*)	0.75	0.12	0.51
Blackgram (*vigna mungo*)	0.85	0.18	0.53
Non edible Oil Cakes			
Castor cake	4.3	1.8	1.3
Cotton cake	3.9	1.8	1.6
Karani cake	3.9	0.9	1.2
Mahua cake	2.5	0.8	1.8
Neem cake	5.2	1.0	1.4
Safflower cake	4.9	1.4	1.2
Edible Oil cakes			
Coconut cake	3.0	1.9	1.8
Groundnut cake	7.3	1.5	1.3
Niger cake	4.7	1.8	1.3
Rape seed cake	5.2	1.8	1.2
Sesame cake	6.2	2.0	1.2

Source: "Handbook of Manures and Fertilizers" 1964.

7.4 Green Manures

Green manuring can be defined as a practice of ploughing or turning into the soil undecomposed green plant tissues for improving physical structure as well as soil fertility. Green manuring, wherever feasible, is the principal supplementary means of adding organic matter to the soil. The green-manure crop supplies organic matter as well as additional nitrogen, particularly if it is a legume crop, due to its ability to fix nitrogen from the air with the help of its root nodule bacteria. The

Green manuring is the practice of turning into the soil undecomposed green plant tissue. The function of a green manure crop is to add organic matter to the soil. As a result of the addition, the nitrogen supply of the soil may be increased and certain nutrients made more readily available, thereby increasing the productivity of the soil.

green-manure crops also exercise a protective action against erosion and leaching. Green manure to be incorporated in soil before flowering stage because they are grown for their green leafy material, which is high in nutrients and protects the soil. Green manures will not break down in to the soil so quickly, but gradually, add some nutrients to the soil for the next crop.

Green manure is prepared by cultivation of some crops and afterward ploughing or turning them into the soil. The practice of ploughing or turning into the soil undecomposed green plant tissue for the purpose of improving physical condition as well as fertility of the soil is referred to as green manuring and the manures obtained by this method is known as green manures. The legumes used as green manuring crops provide nitrogen as well organic matter to soils. The legume crops have ability of acquiring nitrogen from the air with the help of its root nodules bacteria. Example of legume crops such as sunnhemp (*Crotalaira juncea*), Dhaincha (*Sesbania aculeata*), moong (*Phaseolus aureus*), cowpea (*Vigna catjang*), lentil (*Lens esculenta*), berseem (*Trifolium alexandrinum*), Lucern (*Medicago sativa*), pea (*Pisum sativum*) *etc*., which are used as green manuring. The non-leguminous crops such as mustard (*Brassica* sp.), wheat (*Triticum* sp.), radish (*Raphanus sativas*), carrot (*Dancus carota*), jowar (*Sorghum vulgare*), maize (*Zea mays*), sunflower (*Helianthus annus*) *etc*. used as green manuring provides only organic matter to the soil. Green manuring is done in May-June and turning of green manuring is done in July-August. It is done before the cultivation of kharif crop. The seed of green manuring crop such as sunnhemp, dhaincha, cowpea *etc*. is sown in barefallows land and is buried in kharif season. This practice is followed in Uttar Pradesh, Punjab, Bihar, Rajasthan, Madhya Pradesh and West Bengal. The best time for turning the green manuring crop is when it is at flowering stage.

The nutritional potentials and nutritional contents of some important green manures are given in the Tables 7.5 and 7.6 respectively.

Table 7.5. Nutrient Potential of Green Manures

Green Manure	*Biomass (tones)*	*N Accumulobase (kg/ha)*
Sesbania aculeate	22.50	145.00
S. rostrate	20.06	146.00
Crotalaria juncea	18.40	113.00
Tephrosia perpurea	6.80	6.00
Green gram	6.50	60.20
Black gram	5.12	51.20
Cow pea	7.12	63.30

Table 7.6. Nutrient Content of Important Green Manures

Crop	*Nutrient Content (per cent) on Dry Weight Basis*		
	N	*P*	*K*
Green manure			
Sesbania aculeata	3.3	0.7	1.3
Crotalaria juncea	2.6	0.6	2.0
Sesbania speciosa	2.7	0.5	2.2
Tephrosia purpurea	2.4	0.3	0.8
Phaseolus tribolus	2.1	0.5	–
Green leaf manure			
Pongamia glabra	3.2	0.3	1.3
Glyricidea maculeata	2.9	0.5	2.8
Azadirachta indica	2.8	0.3	0.4
Calatropis gigantecum	2.1	0.7	3.6

7.4.1 Advantages of Green Manures

Usage of green leaf manure is advantageous both for crops and soil. The advantages are:

- ✰ As they decompose rapidly, it is easy to retain the organic matter in the soil.
- ✰ Green manures improve both physical and chemical properties of the soil.
- ✰ They provide energy to microbes. . They provide nutrients to the standing crop and also to the next crop.
- ✰ Addition of green manure crops to the soil, acts as much and prevent soil erosion.
- ✰ Leaching of nutrients in light soils can be prevented by addition of green manure.
- ✰ Cultivating green manure crops can control weeds.
- ✰ Majority of green manure crops being legumes, use of nitrogenous fertilizers can be minimized.

7.4.2 Different Green Leaf Manure Crops

Cowpea

Cowpea is one of the important leguminous green leaf manure crops. As this plant is easily decomposable and very well suited for green manure purpose. June-July months are best suited for sowing of this manure. Even though it is being cultivated in summer months (March to April). Use of effective Rhizobium bacteria increase the fixation of nitrogen up to 40 kg/ha.

Dhaincha (*Sesbania aculeate*)

Dhaincha is suitable for loamy and clayey soils. It is fairly resistant to drought as well as stagnation of water. It grows well even in alkaline soils and corrects alkalinity if grown repeatedly for 4-5 years. The roots have plenty of nodules. It yields about 10-15 tonnes of green manure per ha and requires a seed rate of 30-40 kg/ha. Use of effective Rhizobium strain with seeds fixes the Nitrogen 1 kg/day.

Sesbania speciosa

It is a valuable green manure for wetlands and can be grown in a wide range of soils. Seed production is prolific however, pods are frequently attacked by insects. This green manure can be raised on the field borders. Sesbania seedling (21days) can be planted in a single line at 5-10 cm apart in the borders of the fields. In about 90 days it produces about 2-4 tonnes of green manure per ha. It does not affect the rice yield by shading or root effect. If second rice crop is planted immediately after the first crop, the manure can be incorporated into the field. About 300-400g of seeds are sufficient to raise nursery and plant the seedlings around the boundary of one hectare. To control insects *Verticillium lacanii* (Liquid) fungi is useful.

Sunnhemp (*Crotalaria juncea*)

It is a quick growing green manure crop and gets ready for incorporation in about 45 days after sowing. It does not withstand heavy irrigation leading to flooding. The crop is at times subject to complete damage by leaf eating caterpillars. The crop can produce about 8-12 tonnes of green biomass per ha. The seed requirement is 30 kg/ha.

Sesbania rostrata

One of the important features of this green manure is that in addition to the root nodules, it produces nodules in the stem. The stem nodulation is an adaptation for waterlogged situation since flooding limits growth of green manures and may reduce root nodulation. Under normal condition, both root and stem nodules are effective in N fixation. It has higher N content of 3.56 per cent on dry weight basis. Biomass production is higher during summer (April – June) than in winter (Dec. – Jan.) season. This green manure can also be produced by raising seedlings (30 days old) and planted in the paddy field along the bunds or as intercrop with rice. Use of Rhizobium bacteria increase the nitrogen fixation about 60-100 kg/ha/year.

Wild Indigo (*Tephrosia purpurea*)

This is a slow growing green manure crop and cattle do not prefer to graze it. The green manure is suitable for light textured soils, particularly in single crop wetlands. It establishes itself as a self sown crop and the seeds remain viable till the harvest of rice. On an average about 3-4 tonnes of green manure is obtained in one ha. The seed rate is 30 kg/ha. The seeds have a waxy impermeable seed coat and hence scarification is required to induce germination. Soaking seeds in boiling water for 2-3 minutes is also equally effective in promoting germination.

Indigo (*Indigofera tinctoria*)

It resembles wild indigo and is along duration crop with more leafy growth. It comes up well in clayey soils with one or two irrigations.

Pillipesara (*Pheseolus trilobus*)

This is a dual purpose crop yielding good fodder for the cattle and green manure. Pillipesara comes up well in hot season with sufficient soil moisture. Loamy or clayey soils are best suited. After taking one or two cuttings for fodder or light grazing by animals, the crop can be incorporated into the soil. About 5-8 tonnes of manure can be obtained from one ha.

Glyricidia (*Glyricidia maculeata*)

This is a shrubby plant that comes up well in moist situations. Under favourable conditions, it grows well like a tree. It can be easily grown in waste lands, farm road sides, field bunds, *etc.* The crop can be established by stem cuttings or seedlings planted in the field borders. It can be pruned for its tender loppings and compound leaves for green leaf manuring at the time of puddling rice. On an average, a well-established plant yields 12-15 kg green matter. About 400 plants on the peripheral bunds yields 5-6 tonnes green manure/ha.

Karanj (*Pongamia glabra*)

It is a leguminous tree grown in wastelands. On an average, a tree can yield 100-120kg of green matter. The leaves contain about 3.7 per cent N (on dry weight basis).

Calatropis (*Calotropis gigantca*)

On roadsides and fallow lands, the plant grows wild under different soil and climatic conditions. The leaves are more succulent and a plant can produce about 4-5 kg of green matter. Besides it also helps in controlling soil born pests like termite.

7.5 Compost

Composting is the natural process of recycling organic matter, such as leaves and food scraps, into a valuable fertilizer that can enrich soil and plants. Anything that grows decomposes eventually; composting simply speeds up the process by providing an ideal environment for bacteria, fungi, and other decomposing organisms (such as worms, sowbugs, and nematodes) to do their work. The resulting decomposed matter, which often ends up looking like fertile garden soil, is called compost. Fondly referred to by farmers as "black gold," compost is rich in nutrients and can be used for gardening, horticulture, and agriculture.

Large quantity of fresh crop residues, on application directly to soil, causes extremely severe nitrogen immobilization and development of excessive reduced condition in the soil. To overcome such problems organic residues are piled up, moistened, turned occasionally to aerate and allowed adequate time to decompose partially and bring down the carbon nitrogen ratio to about 30. This process is called composting. Compost, is utilized for improving or maintaining soil fertility. The collected organic refuse may be of rural and urban origin and may include straw, leaves, paddy husk, ground nut husk, sugarcane trash, bagasse, cattle dung, urine, crop residues, city garbage, night soil, sewage, kitchen and vegetable wastes, hedge clippings, water hyacinth and all other residues counting organic matter. During a under thermophilic and mesophilic conditions in windrows, heaps or pits adequate moisture and aeration are essential. The final product is brown to black colored humified material which on addition to soil replenishes plant nutrients, maintains soil organic matter content and helps in improving the physical, chemical and biological conditions of the soil.

Organic discards can be processed in industrial-scale composting facilities, in smaller-scale community composting systems, and in anaerobic digesters, among other options. This guide focuses primarily on home composting, which is a great way to keep your organic discards out of the waste stream and produce a valuable soil amendment for your own use.

Composting is organic manure artificially prepared from plant residues and animals waste products. The process of making compost is known as composting. It is largely a biological process in which aerobic and anaerobic microorganism decomposed organic matter and lower the C: N ratio of the refuse. Village or rural compost is prepared from farm waste like straw, crop stubbles, crop residue such as sugarcane trash, cowshed and hedge clippings. This type compost contains 0.4-0.8 per cent N, 0.3-0.6 per cent P2O5 and 0.7-1.0 per cent K_2O. Town or urban compost is prepared from town waste and night-soil. This type compost contains 1.0-2.0 per cent N, 1.0 per cent P_2O_5 and 1.5 per cent K_2O.

7.5.1 Beneficial Effect of Compost on Soils

Compost improves the physical, chemical and biological properties of soils. Sandy soils are compacted and clay soils become loose. The water holding and heat absorbing capacity of soils increase and thus compost improves the permeability of soils. The alkaline and saline soils are also made less deleterious making plant growth easier in such soils. Compost are made from plant refuse which contains all the nutrients needed by plant are made available to plant, when compost are added to soils. The nutrients lost from the soils are again returned to the soils for future use by plants. Besides this, compost contains these nutrients specially N, P, K and Ca in organic combinations and when it decomposes, these nutrients are liberated in available form for the use of plants. Compost is accelerating the microbial activities in different process like ammonification, nitrification and nitrogen fixation. Compost especially activates those fungi which live in close association with the roots of plants and trees, *e.g.* mycorrhiza. They play important role in transferring certain essential plant nutrients and accessory food from soil to the plant. Compost is supposed to contain small amount of growth promoting substances which are known as hormones.

Vermicompost prepared by using of earthworms. Earthworms consume decomposed solid waste of plant and animal origin; convert them to valuable biofertilizer by dropping/casts. Vermicompost is superior type of compost manures which contain 1.0-1.5 per cent N, 1.8-2.2 per cent P2O5 and 1.0-1.5 per cent K2O. It is nowadays regarded as a very important of the organic farming package. It is easy to prepare, has excellent properties and is absolutely harmless to plants. It is rich in macro and micronutrients, vitamin, growth hormones and immobilized microflora. Application of vermicompost facilitates easy availability of essential plant nutrients to plants.

7.5.2 Requirements of Composting

In general, composting is carried out in open pit or above ground, by filing alternate layers of organic wastes and other materials including top soil, cattle dung, half decompose farm yard manure, rock phosphate and other amendments. If the organic wastes are largely high-carbohydrate materials, some fertilizer nitrogen is needed. The addition of poultry waste and farm yard manure while layering the compost pit, tends to speed up decomposition and helps to improve the texture of the product. The optimum C:N ratio of the composting materials is below 40. Good aeration in the compost pile is essential. It is good to mix succulent organic materials with the materials that decompose slowly. This prevents packing into soggy anaerobic mixtures. Since composting is a biological process, sufficient moisture for the proper development of microorganisms is essential. The materials should not be too dry or soggy. The requirement of moisture for microorganisms is almost similar to that of higher plants. The optimum moisture content of the composting materials has been found to be 60 per cent of the total water holding capacity of the substrate.

7.5.3 Microbiology of Composting

At the initial stage, the easily degradable organic matters like carbohydrates, fats, proteins get degraded by the action of mesophilic fungi. Due to its action, some amount of heat energy is formed hence, the temperature of the composting substrates is 400 C or less, mesophilic fungi and acid producing bacteria appear. The proportion of the three groups of organisms — bacteria, fungi and actinomycetes is related to dominance of organic constituents., Water soluble simple sugars encourage rapid bacterial proliferation while starch benefits the actinomycetes in particular substrates rich in proteins or amino acids stimulate the spore forming bacilli. Details of microbial genera capable of utilising cellulose, hemicelluloses, starch, lignin *etc.* may be seen in Table 7.5. At the end of mesophilic action, the thermophilic fungi/bacteria will start its action and degrade some part of organic matters. Hence the temperature of 65^0 C to 70^0 C attained during aerobic decomposition in compost pit leads to the destruction of most of the pathogens, parasites and weed seeds present in the original material. In addition to the effect of higher temperature, some pathogens and parasites are also killed due to their failure to withstand the competition with other microorganisms. The Actinomycetes and spore forming bacteria start its action to decompose cellulose and Hemicellulose present in the waste. During anaerobic decomposition as prevailed in biogas plant and Bangalore method of composting, temperature does not rise to the extent lethal to parasites in a relatively short time. When sewage sludge or night soil is composted, anerobic decomposition should be proceded by aerobic composting atleast for a week. The natural death of pathogens and parasites occurs under anaerobic environment and the microbial antagonism eventually eliminates them in relatively longer period of six months. After decomposition of maximum amount of organic matters, the temperature of the medium will slowly decreases. During the temperature from 60^0 c to 40^0 c again, the thermophilic fungi start to decompose remaining part of organic matter. After this, the temperature again decreases further. The white colour fungi commonly appears on the waste that shows presence of mesophilic fungi.

During this fourth phase, the temperature comes down to amount 30-40^0 c. In this stage, again mesophilic fungi start its action and degrade some amount of lignin. After this, *i.e.* in final stage of this composting process, the temperature remains constant. Table 7.7 shows organisms in compost.

Table 7.7. Organisms in Compost

Microflora/Microfauna	*Organisms*	*Number/g of Moist Compost*
Microflora	Bacteria	10^8-10^9
	Actinomycetes	10^5-10^8
	Fungi, moulds and yeasts	10^4-10^6
Microfauna	Protozoa	10^4-10^5
Microflora	Fungi (Mushrooms and Toadstools)	
Macrofauna	Mites ants Termites Spring tails	
	Millipedes Centipedes Earthworms	

Source: FAO Soil Bulletin (1987).

7.5.4 Factors Influencing the Bio Composting Process

- **Particle Size:** If particle is small, the space for the growth of microorganisms will be more which ultimately increases the microbial activity and fastened the composting process.
- **Nutrients:** Any waste materials which do not contain heavy metals/toxic waste should be used. The list of nutrient value as in Table 7.5 should be

taken in account to increase the nutrition value of the waste. Preferably, the local waste with some available agro-waste is always ideal.

- ☆ **Moisture:** The total net weight moisture should be maintained 50-60 per cent which is optimum for microbial growth. Hence, this percentage of moisture should be maintained throughout process which will increase the process of decomposition. If the moisture is below 50 per cent, the microbial activity is less, which results slow down the composting process. If the moisture is above 60 per cent, the anaerobic condition (clogging) will take place which also slows down the activity of microorganisms.
- ☆ **Aeration and Agitation**: For the growth and activity of microorganism, oxygen supply is most essential part. Any process to be adopted only to ensure the oxygen supply. The agitation, turning by any tools or by means of Aero-tilling may be used.

7.6 Composting Process

Bio-composting is a process for the rapid conversion of organic waste into a thoroughly decomposed, stable and humus rich compost for used as a fertilizer and soil conditioner. The process is aerobic and the technology highlights the following activities for large and commercial scale Bio-composting.

- ☆ **Windrowing:** The waste materials should be chopped in small size and formed in windows of 3 meters width and 1.2 meters height, the length according to the availability of the land.
- ☆ **Inoculation:** Mixed population of microorganisms are sprinkled over the windrows at the rate of 4kg/tone of waste materials. In case of liquid 2 litres/tone is enough as the population of fungus and bacteria is 100 times more than solid base inoculum.
- ☆ **Aero tilling:** The windrow is aero tilled in alternative days by using special type of machine called "Aero-tiller" or manual method. It helps for uniform mixing and provides oxygen to the microorganisms.
- ☆ **Application of Spent wash:** To maintain the optimum moisture of 50-60 per cent and to maintain temperature between 65-700 c for high rate composting, the nutrient rich spent wash is sprayed on the windrows, if it is available. Otherwise, any wash like cow dung wash, vermiwash, kitchen wash, any animal wash *etc.* can be sprayed.

Composting Procedure

A composting cycle takes 11 weeks to complete and involves the following activities

I[st] Week

1. Collection/Hauling of waste materials.
2. Formation of Windrows and Trimming.

3. Aero-tilling-for uniform mixing of material and to bring down the moisture at optimum level.
4. Inoculation of microorganisms in windrows only high temperature tolerate fungus/bacteria should be used.

2nd to 8th Week

Application of wash and Aero tilling Monitoring moisture content and temperature of windrows (by any thermometer)

9th to 11th Week

After 11th week, the ready Biocompost is enriched with Biofertilizers.

Enrichment of compost with Biofertilizers: N-fixing bacteria-Azotobacter, Azospirillum each 2 kg/tone if solid, 1 litre/tone if liquid.

P-solubilizing bacteria-*Bacillus polymixa etc*-4kg/ton if solid, 2 litres/ton,1 litre/tonne if liquid.

K-Mobilizing bacteria-*Fraturia aurantia* 4kg/ton if solid, 2 ltrs/ton if liquid.

Enrichment of Compost with Bio-Agents

N-fixing bacteria-Azotobacter, Azospirillum each 2 kg/tone if solid, 1 litre/tone if liquid.

P-solubilizing bacteria-*Bacillus polymixa etc*-4kg/ton if solid, 2 litres/ton,1 litre/tonne if liquid.

K-Mobilizing bacteria-*Fraturia aurantia* 4kg/ton if solid, 2 ltrs/ton if liquid.

Enrichment of Compost with Bio-Agents

Bio-Control agents like *Trichoderma viride, Pseudomonas* Fluorescence at the rate of 2kg/tone of each if solid, 500ml/tone if liquid form.

7.7 Old Methods of Compost Production

7.7.1 Chimney and Wall Method

For wall aeration technique two brick walls, 30 cm apart, 1 m high, 23 thick and having 40 holes of size 22 cm x 10 cm each are constructed in the centre of a 2 x 3 m pit. Chimney aeration is provided by constructing two one meter high chimneys, one metre apart on rectangular base of 23 cm. Each chimney has 40 holes of similar size as provided in wall aeration. Substrates are filled in layers with each layer consisting of three sublayers. In a trial carried out at Jabalpur, the sublayers were consisted of (1) overnight water soaked paddy straw 50 kg (air dry basis) (2) dung slurry (20 kg dung in 40 litres water) (3) mixture of 20 kg well pulverised soil, half kg urea and 1.5 kg urea and 1.5 kg rock phosphates (1.5 per cent available phosphorus, 150 mesh passed) Total quantity of substrate in ten such layers in each pit was one tonne. After one month, the towers and chimney were sealed with dung and mud mixture. Chimney aeration yielded 60 per cent recovery with nitrogen ranging form 1.10 to 1.70 per cent. Further trials have shown that the biomass of obnoxious weed, *Parthenium histophorus* yields better quality compost than rice straw. However, before using the parthenium compost, it is to be ensured that the weed are not viable. Enrichment of compost with Nitrogen Phosphorus and Potassium can be achieved by Azotobacter, Azospirillum and phosphate solubilizing and potassium mobilizer Biofertilizers after the thermophilic phase is subsided at the rate of 2-4 kg each per tonne of substrate. The biofertilisers may be dissolved in 50 litres water and poured in holes previously made in compost piles. If biofertilisers are not available while composting is in progress, enrichment can be done by mixing biofertilisers with the harvested compost and heaping the treated compost in shed for at least two weeks. By doing so, increase in the population of Azotobacter, Azospirillum PSM,KMB has been noted. Inoculation of the substrates with celluloytic and lignolytic microorganisms like *Trichoderma harzianum, Aspergillus niger, Aterreus etc.* has been found to accelerate the decomposition during composting. Likewise, *Trichoderma harzianum* is the msot effective organism for making compost from rice straw.

7.7.2 NADEP Method

NADEP method of compost making has been developed by a farmer, Narayan Rao Pandhari Pande, in Maharastra, India. This method is based on the principle of aerobic decomposition with natural flow of optimum air. The substrate is converted at the top by plastering with dung and soil to minimise the loss of moisture. To obtain 2-2.5 tonnes of compost with 0.6-1.0, 0.5-0.8 and 1.2-1.5 per cent nitrogen, phosphorus and potash respectively, the required raw materials are 1.4 to 1.5 tonnes or organic refuse, 90-1000 kg of cattle dung, 1.7 to 1.8 tonnes of pulverised dry soil and 1500-2000 litres of water. These materials are filled layer by layer in a tank of 3 m x 2 m x 1 m size and made up bricks with holes (15-20 cm rectangular) on all four walls for easy entry and circulation of air. The tank is constructed above ground at high lying area to avoid entry of rainwater from surrounding place. Internal surface of tank is painted with dung slurry.

The tank is filled in layers. The first layer is made by spreading 100-110 kg of organic wastes on the floor of the tank. This is followed by a second layer containing a slurry of 4-5 kg cattle dung in 125-150 litres water. Over the second layer, 50-60 kg of pulverised good quality soil is spread. The three layer combination is repeated till the tank is filled upto 45-50 cm above the brick level. The complete filling is done within two days, using materials of 10-11 layers. Topping of the tank is done by 5 cm plastering with a paste of dung and soil. Cracks are not allowed to develop on the heap to prevent the gas leakage. After 15-20 days, when the substrate shrink down, a second filling is made in a way similar to that adopted in the beginning. The top is then arranged in a hut like shape and replastered. The moisture level of the mass is maintained at 15-20 per cent by sprinkling with water and dung slurry through holes. Normally the substrate takes 3-4 months to attain maturity without turning. From one tank about 4.5 - 5.0 cubic metre mature compost 1.5

- 2.0 cubic metre under composed raw refuse, weighing about there tonnes are obtained. According to an estimate, from the annual collection of dung from one cow, 80 tonnes of compost can be prepared which contain 800 kg nitrogen, 560 kg phosphorus and 1000 kg potash (Bonbatkar, 1989).There are some limitations in using this method of composting like unavailability of dry soil during rainy seasons and water during summer. The ratio of organic refuse and soil is about 1:1. Soil is added to provide microorganisms for carrying out decomposition and nitrogen fixation. The compost harvested does not give dark brown colour, has a higher density and lower nutrient composition as compared to the product obtained from the chimney and tower aeration method.

7.7.3 Padegaon Method

This method is recommended for composting resistant substrates like sugarcane trash and cotton stubbles. These materials are shredded into 30 cm size particles and trampled to make a 30 cm thick above ground layer. This layer is drenched with a slurry consisting of wood ash, cow-dung and soil. Four or five such layers are added to the pile. The completed heap is about 1.5 m high, 2m wide and as long as necessary. Since the material is very resistant to decay, the heap is turned each month, re-trampled and sufficient water is added to keep it moist. The material is ready for use in about five months. The compost compares very well in composition with farm yard manure.

7.7.4 Indore Method

Sir Albert Howard (1924-26) at Indore, Madhya Pradesh, developed this method in which the conservation of cattle urine is effected by getting it absorbed in rice straw, straw dust and other organic wastes used as bedding in cattle shed. The urine soaked material along with fresh cow dung serves as major source of nitrogen for the microorganisms involved in composting. The material collected form cattle shed is spread evenly in a pit to form a layer of 10-15 cm thick. To this layer is added dung slurry made of 4.5 kg dung + 3.5 kg urine earth + 4.5 kgs inoculum from 15 day old compost pit. Water is then sprinkled to achieve 100 per cent saturation. The layering is repeated to fill the pit within seven days. The material is turned three times, first two turnings in 15 days interval after filling the pit and third turning after one month of the first turning. To provide succulent biomass, seeds of sunhemp are grown on compost heaps and at the first turning, the green plants are turned in. During rainy season the piling of 20 cm carbonaceous material (leaves, hay, straw, saw dust, wood chips, corn stalks *etc.*) and 10 cm nitrogenous materials (fresh grass, weeds, digested sewage, sludge, poultry litter) in alternate layers is repeated until the pile is one metre high. The recommended size of the heap is 2.4 m square at the base and 2.1 m square at the top. The method is highly labour intensive and less suitable to those farmers who do not have enough cattle and irrigation facilities. The method being aerobic in nature, it hastens the maturity period and results in substantial loss of organic matter and nitrogen. Table

8 and 9 shows some bacteria,fungi and actinomycetes capable of utilizing various components of organic matter.

Table 7.8. Some Bacteria Capable of Utilizing Various Components of Organic Matter

Substance	*Bacteria*
Hemi cytophaga	*Achormobacter, Bacillus cellulomonas, Cellvibrio, Clostridium, Cytophaga, Pseudomonas, Sporocylophaga, Vibrio bacillus, achromobacter, cellulose pseudomonas, Sporocytophaga, Lactobacillus, Vibrio*
Starch	*Achromobacter, Bacillus, Chromobacterium Clostridium, Cytophaga, Rhizopus, Flavobacterium, Micrococcus, Pseudomonas, Sarcina, Serratia,*
Hydrocarbons and Pesticides	*Methanobacterium, Methanobacillus, Methanosarcina, Methanococcus, Pseudomonas*
Proteins	*Pseudomonas, Corynebacterium, Mycobacterium, Bacillus, Vibrio*

Table 7.9. Some Fungi, Actinomycetes Capable of Utilising Various Components of Organic Matter

Substance	*Fungi*	*Actinomycetes*
Cellulose	*Alternaria, Apergillus, Fusarium Penicillium, Polyporus, Rhizoctonia, Rhizopus, Trichoderma*	*Micromonospora, Nocardia, Streptomyces, Streptosporangium*
Hemicellulose	*Alernaria, fusarium, truchithecium aspergullus corolus, rhizopus, Rhizopus, Zygorhynchus, Chaetomium Helminthosprium Penicillium Fomes polyporus*	*Actinomycetes, Streptomyces*
Starch	*Asoergillus, Fomes, Fusarium, Poyporus*	*Micrimonospora, Nocardia*
Hydrocarbons and pesticides proteins	*Streptomyces Streptomyces Nocardia Asdpergillus Nocardia Streptomyces*	

7.7.5 Bangalore Method

In this method, the disadvantages of Indore method are overcome by slowing down the rate of decomposition and avoiding the turnings. The substrates usually composted in this method are town refuse and night soil which are spread in alternate layers of 15 cm and 5 cm in trenches or pits. When the pit is filled to 15 cm above the ground level, it is sealed to prevent loss of moisture. After the initial aerobic decomposition for 8 - 10 days the material undergoes semi anaerobic decomposition. During this stage the rate of decomposition slows down taking about 6 - 8 months for the compost to be ready. Often, the composting period is more than eight months due to high carbon: nitrogen ratio. Loss of organic matter and nitrogen is negligible and percentage recovery of compost is more. But, this method is not adaptable to heavy rainfall areas.

7.8 Farm Yard Manure (FYM)

FYM is partially composed dung, urine, bedding and straw. Dung comes mostly as undigested material and the urine from the digested material. More than 50 per cent of the organic matter that is present in dung is in the form of complex products consists of lignin and protein which are resistant to further decomposition and therefore the nutrients present in dung are released very slowly. The nutrients from urine, becomes readily available. Dung contains about 50 per cent of the nitrogen, 15 per cent of potash and almost all of the phosphorus that is excreted by animals. Straw, saw dust or other bedding materials are used in cattle sheds to reduce the loss of urine and to increase the bulk of manure. On an average, about 3 - 5 kg bedding material per animal is used by farmers. FYM contains approximately 5 - 6 kg nitrogen, 1.2 - 2.0 kg phosphorus and 5 - 6 kg potash per tonne. The quantity and quality of FYM depend upon the type (draught, mulch) and age of the animals, the way they are feed and the care taken to collect and store the material. Though FYM is the most common organic manure in India, the farmer, in general, do not give adequate attention to the proper conservation and efficient use of the resource. For preparing better quality FYM, the use of pit method for areas with less than 1000 mm precipitation and heap method for other places is recommended. In the pit method, the cattle shed wastes are conserved in pits of 2 m wide, 1 m deep and of convenient length with a sloping bottom towards one end. In the pit an absorbent layer is created at the bottom by spreading straw at the rate of 3 - 5 kg per animal kept. The substrate containing well mixed dung, urine and straw is spread over the absorbent layer daily to form a layer of 30 cm thick and the process continued until the pit is filled. Each day's layer should be pressed, moistened if dry and covered with a 3 - 5 cm layer of well ground fertile soil to hasten the decomposition and to absorb the ammonia. The pit should be prepared on high lying area to avoid the entry of rain water. In the heap method, the daily collections from cattle shed are spread in uniform layers until the heap attains a maximum height of one meter above ground. The top of the heap is rounded and plastered with dung and mud mixture. In both the pit and heap methods aeration is allowed in the beginning and later on anaerobic conditions set in and continue for a long period. The manure is ready for use after 5 - 6 months. These methods should be initiated prior to rainy season and continued throughout the year. If properly preserved, the quantity of manure that can be produced per animal per year would be as much as four to five tonnes containing 0.5 per cent nitrogen. This is in contrast to one or two tonnes per animal per year containing 0.5 per cent nitrogen, that is obtained by indigenous method. The materials should not contain any heavy metal.

7.9 Urban Compost, Sewage and Sludge

Metropolitian areas are facing major problem in the disposal of large volumes of liquid and solid wastes generated by urban and industrial activities. In India the urban population of 240 million produces approximately 29 million tonnes of urban solid wastes (USW) annually and 1010 litres of liquid wastes per day. Broadly liquid wastes constituent sewage and sludge. Sewage is spent water consisting of water carried wastes from commercial and industrial areas and surface and ground water that enter the sewage system. The sewage and sludge are useful sources of water for irrigation and nutrients and the urban solid wastes can serve as nutrient source. The urban solid wastes presently available in India have the potential to supply 145, 70 thousand tonnes of nitrogen, phosphorus and potash.

Urban composting methods are either-non-mechanical or mechanical. Mechanical composting (anaerobic) of city wastes is a natural process and in practice several months are required for the transformation of wastes into compost. The amount of heat generated in small and the required temperatures are not obtained for the destruction of pathogenic organism. For the reasons that anaerobic composting liberates bad odors, is of long duration and is incapable of destroying pathogenic organisms, it is not generally adopted.

7.10 Composting of Urban Refuse

7.10.1 Sewage and Sludge

Sewage effluent treatment is based on biological decomposition of organic matter present in it. Primary treatment includes removal of solid matter by physical means such as screening and sedimentation. It is possible to remove 30- 50 per cent of the suspended solids. Secondary treatment is done through activated sludge process. In this process, the sewage is aerated either by diffusion or mechanical means and the aerobic microbial decomposition process is activated. The sludge is separated from the effluent in a settling tank after a suitable interval. An overall reduction of 85-95 per cent of suspended solids and BOD and 90-98 per cent reduction in bacterial are achieved through the use of this method combined with the primary treatment. The resultant effluent usually has been suspended solid content of 20 ppm which is well within the safe limit. Bureau of Indian Standards (BIS) has laid down tolerance limits for the effluents to be discharged on land for irrigation and the effluents to be discharged into public sewers.

7.10.2 Composting Sludge

Sludge collected from sewage treatment process are composted with other organic materials for use as manure. While composting, most of the nitrogen in the sludge is assimilated by the organisms and toxic substances are diluted. Methods for composting de-watered sludge depend upon its moisture content. If the sludge contains less than 60 per cent moisture, the use of organic absorbent to adjust the moisture content is not necessary. In fact air-dried lime sludge (55-60 per cent moisture) and the mixture of de-watered sludge (70 per cent moisture) and well composted sludge (30 per cent moisture), whose moisture content does not exceed 55-60 per cent are successfully stabilized under aerobic conditions for 7-10 days (Date, 1982). If the mixture of sludge and absorbent organic material containing less than 60 per cent moisture the sludge can be stabilized without any difficulty. But, when slowly decomposing absorbent's such as rice husk and saw dust are used the primary product must be stockpiled and matured for 1-2 months to accelerate decomposition of absorbents.

Activated sludge decomposes relatively easily in soil, and thus most of the nitrogen is mineralized within one month. It has been estimated that one tonne of sewage sludge provides 46.9 kg nitrogen, phosphorus and potash but has the potential for heavy metal build up with continued application and resultant human health problems. Heavy metals like cadmium, lead, nickel, chromium *etc.* are generally not found in compost, FYM and city refuse in amounts that will be toxic to the soil, plants and mankind.

Table 7.10. Chemical Characteristics of Bio-gas Slurry, Sewage Sludge, Pressmud and City Refuse

Parameters	*Biogas Slurry*	*Sewage Sludge**	*Pressmud*	*City** Refuse*
Total N, per cent	1.40-1.84	1.08-2.34	1.1-1.19	0.58-0.61
Total P_2O_5 per cent	1.10-1.72	0.84-2.14	2.12-2.43	0.59-0.71
Total K_2O per cent	0.84-1.34	0.53-1.73	1.98-2.03	0.67-0.73
Organic Carbon per cent	35.00-38.40	19.50-31.80	23.50-26.70	12.79-16.04
Zinc, mg/kg	103.00-115.60	700.4-1245.70	237.50-285.40	-
Copper, mg/kg	50.60-67.50	193.7-535.50	112.40-131.70	-
Manganese mg/kg	231-00-294.70	176.2-46520	-	-

7.10.3 Composting of Urban Refuse

Before composting, non-organic and resistant materials like rags, plastics, glasses iron nails, tin, aluminium foils *etc.* are separated from the substrate. This is followed by shredding the organic components to particles of desired size (5 cm) and keeping the moisture level at 40 - 60 per cent. Aeration is promoted by turning the material during composting. Turning also facilitates the development of uniform temperature over the entire mass and subjects the material to high temperature.

The frequency of turning and the composting mass depends upon the type of refuse and moisture content. The schedule normally followed is: 1st day stacking and adding of moisture, 5th and 10th day turning and spraying of moisture, 15th day — turning, and 20th day removal to process mill for screening. During rainy season frequency of turning is increased to promote aerobic condition in the piles. Fly control can be easily managed by not allowing the refuse to remain for long in the open but placed in stacks immediately for composting where the high temperature and environmental conditions are not favourable for the continued emergence and breeding of flees.

The carbon: nitrogen ratio of garbage available in India shows a range of 20 to 30 and it is possible to get stabilized compost in about 20 days (Jain, 1980). Mechanical composting is aerobic which provides conditions for rapid conversion of organic wastes into compost. Non-mechanical method of composting is employed when volume of refuse to be handled is small. In this method readily computable materials, such as night soil, animal wastes, sewage – sludge and garbage and relatively resistant organic matter, such as straw, leaves and municipal refuse are piled in alternate layers on ground or in pits. The composting mass is turned regularly for 8 - 12 weeks and stirred on ground for a further period of 4 - 6 weeks. After about 5-8 turnings and a total period of four months, the compost is ready for use.

Chapter 8

Vermicomposting

8.1 Introduction: An Overview

Vermiculture is the culturing of earthworms. Recent years, much interest in vermiculture has been paid by the large producers and farmers as earthworms play a major role in soil improvement, organic matter decomposition and in enhancing the quality of agricultural produce.

Vermicomposting is applied for composting various non-toxic organic solids and liquid wastes available from cities, dairies, sugar and distillery units, pulp and paper mills, tanneries, fermentation industries and food processing units.

Earthworms are used as biodegraders for composting such refuse, after the thermophilic activities have subsided. In vegetable and plantation crops where mulching with organic debris is a common practice, application of vermicastings at the rate of 5 tonnes/ha as the first layer is recommended.

Vermicastings consisting of excreta of earthworms and the cocoons released by them are rich in organic matter and plant nutrients. In the presence of vermicasts the declay of orgnic refuses and the formation of compost is accelerated. The process could be continued uninterruptedly by the addition of dung and other organic wastes at regular intervals to serve as food source for the earthworms.

8.2 Earthworms

Earthworm can generally be called as biological indicators of soil fertility for soil with earthworm most definitely support healthy populations of bacteria, fungi, actinomycetes, protozoans insects, millipedes and other hosts are essential for sustaining a healthy soil.

Earthworms are the unheralded soldiers of the soil. They aerate, break up soil for easier access by plant roots, and help the soil hold more water clean-up dead organic matter by eating it and turning it into world's best plant food. Earthworms are very sensitive to hydrogen ion concentration. Most of the species of earthworms prefer soil with about neutral pH. They avoid drought and dry soils either by migrating to lower layers or by entering a stage of dispose.

The earthworms can tolerate temperature ranging from 0°C to 40°C with pH of 7 but the regeneration capacity is more at 25°C to 40°C and 40-45 per cent moisture contents and partially decomposed organic matter is rich in nitrogen Generally earthworms are more active in moist soil than dry soil. The kind and amount of food material available in soil influence the size of earthworm's population, species diversity, growth rate and cocoon production.

India has about 3000species of earthworms which are adapted to a range of vermiculture needs. They prefer these factor like- soil texture, substrate aeration, temperature, moisture, pH, inorganic salts, organic matter, reproductive potentials, enzymatic digestion and also microbial decomposition of substrate in intestine of earthworms. There are many species of earthworms and each has different preferences for soil conditions.

- ✰ Earthworms are tubular bioreactors, taking in organic waste, processing it with the help of gut micro flora and excreting the vermicasts, the effective biofertilizers for the soil.
- ✰ They are isothermal bioreactors. All the bio-process are sensitive to high temperatures. Earthworms have novel temperature regulating mechanisms.

The most common types of earthworms used for vermicomposting are brandling worms (Eisenia foetida) and Redworms or Red wigglers (Lumbricus rubellus). Often found in aged manure piles, they generally have alternating red and buff-colored stripes. They are not to be confused with the common garden or field earthworm (Allolobophora caliginosa and other species). Although the garden earthworm occasionally feeds on the bottom of a compost pile, they prefer ordinary soil. An acre of land can have as many as 500,000 earthworms, which can recycle as much as 5 tons of soil or more per year. Redworms and brandling worms, however, prefer the compost or manure environment. Passing through the gut of the earthworm, recycled organic wastes are excreted as castings, or worm manure, an organic material rich in nutrients that looks like fine-textured soil. Finished vermicompost should have a rich, earthly smell if properly processed by worms. Vermicompost can be used in potting soil mixes for house plants and as a top dressing for lawns. Screened vermicompost combined with potting soil mixes make an excellent medium for starting young seedlings. Vermicompost also makes an excellent mulch and soil conditioner for the home garden. The survival, biomass production and reproduction of earthworms are the best indicator to evaluate the vermicomposting process.

- They maintain a stable pH throughout their guts. All enzymes are very active in a very narrow pH range.
- They can separate oxygen from the air and supply to the gut micro flora, thus encouraging various aerobic waste-stabilization process and destroying soil pathogens.
- Earthworms build up nitrogen fixing activity in the soil by providing ideal conditions of food, moisture and air to the N-fixing bacteria. Hence crops can flourish without any synthetic nitrogen addition. Relatively small P requirements are met by increased ability of phosphates produces by earthworms.

8.3 Anatomy of Earthworms

The earthworm has a long, rounded body with a pointed head and slightly flattened posterior. Rings that surround the moist, soft body allow the earthworm to twist and turn, especially since it has no backbone. Earthworms are a hermaphrodite, which means they have both male and female sex organs, but they require another earthworm to mate. The wide band (clitellum) that surrounds a mature breeding earthworm secretes mucus (albumin) after mating. Sperm from another worm is stored in sacs. As the mucus slides over the worm, it encases the sperm and eggs inside. After slipping free from the worm, both ends seal, forming a lemon-shape cocoon approximately 1/8 inch long. Two or more baby worms will hatch from one end of the cocoon in approximately 3 weeks. Baby worms are whitish to almost transparent and are 1/2 to 1 inch long. Red worms take 4 to 6 weeks to become sexually mature.

8.3.1 Classification

Earthworms are closely related with their anatomical features. There are 6 types of earthworms namely- *Polypheretima elongate, Lampito mauritii, Pontoscolex corethrurus, Perionxyx excavatus, Octochaetoides beatrix* and *Drawida barwelli* exhibited definite pattern of distribution, soil salinity and temperature tolerance range determines habitat niche of earthworms.

8.3.2 Species Occurrence

- *Polypheretima elongata:* They found at a depth of 4-30 cm low grasslands. They require rich supply of water high salinity and high level of organic matter for survival.
- *Pontoscolex corethrurus*: They found in sub-surface soils which contain very low salinity and high organic matter of shady regions of garden soil.
- *Peionyx excavatus:* Found near cow sheds where compost heaps and is a highly adaptable earthworm to any kind of habitat.

- ☆ *Lampoon mauritii:* They have greater polytrophic adaptation and tolerance to changing micro habitats.
- ☆ *Pontoscolex corethrurus:* Cast of moderate size on soil surface.

8.4 Type of Earthworm Suitable for Vermicomposting

There are 3 types of ecological strategies of earthworms:

8.4.1 Epigeic (Litter Dweller)

Epigeic worms live deep in the soil from 10-30 cm and feed on the humic materials and mineral matter, esophageal; detritivore, lives in and consume litter; small size, uniformly pigmented.

8.4.2 Endogeic (Subsoil Dweller)

Microphage; geophage; (epiendogeic or hypoendogeic)lives in horizontal, branching burrows in organo-mineral layer, consume soil, small to large in size, weakly pigmented.

8.4.3 Anecic (Top Soil Dweller)

Anecic worms can go very deep in to soil up to 60-90 cm and form complicated burrows for their movements. The external abiotic parameters and the poor soil nutrients appear to be controlling factors for growth of earthworm population, macrophage; detritivore, lives in deep vertical burrows, casting on surface; emerges at night to draw down organic matter (plant residue), large as adults (200-1,100mm) brown pigment interiorly and dorsally.

Table 8.1. The Nutrient Status of Vermicompost by *Eisenia foetida*

Sl.No.	*Parameters*	*Eisenia foetida*
1.	pH	7.0
2.	Organic Carbon	27.43
3.	Total Nitrogen per cent	0.60
4.	Total Phosphorus per cent	1.34
5.	Total Potassium per cent	0.40
6.	**C : N Ratio**	**45:70**

The species *Eisenia foetida* (red worm) are the most popular precisely because of the ease replicating the environmental conditions. Perfectly suited to indoor existence and the culturing of this worm requires only minimum of effort, and presenting no hardship for their place of residence. The concentrations of nitrogen, calcium, magnesium, and phosphates are all higher in soil that has earthworms. The movement of the earthworms through the soil loosens it up and makes it valuable to all gardeners and to many others. The enriched soil is naturally processed by the earthworms.

8.5 Vermicomposting Material

The biologically degradable and decomposable organic wastes commonly used as composting material in vermicomposting are as follows:

- Animal dung- cattle dung, sheep dung, horse dung, goat and poultry dropping *etc* may be used for this purpose.
- Agriculture waste- The leaves, stem, husk,peels,vegetable waste, orchard leaf litter, processed food wastes, sugarcane rash and bagasse are all waste used as a agriculture waste.
- Forestry wastes- These are plant products such as wood shavings, peels, saw dust and pulp. All these besides various types of forest leaf litter can be used. The unutilized forest waste such as leaf litter may also be used for vermicomposting.
- City leaf litter- The burnt leaf litter from avenue or residential areas may be used. The leaf litter of mango, guava, grasses and certain weeds (free from seeds)may be used.
- Waste paper and cotton cloth- These are decomposable organic waste. These if not being recycled for other useful products, can be recycled with vermicomposting.
- City refuge-City refuge or garbage on daily production basis comprise important items of city factors and considerable portion of city refuge can be sorted and recycled or composted.

Most of household as kitchen waste with little manipulation can be used for vermicomposting.

- Industrial wastes- wastes like waste from food processing, distillery waste can also be used in vermicomposting.
- Biogas slurry-After recovery of biogas, if not required for agriculture use in convention composting can be used for vermicomposting.

Millions of tons of food waste are buried or burned each year at considerable financial and environmental cost. Instead of discarding the food scraps, can recycle them with the help of worms. Vermicomposting (worm composting) turns many types of organic waste into a nutritious soil for plants. When worm compost is added to soil, it boosts the nutrients available to plants and enhances soil structure and drainage.

Composting food waste offers several advantages:

- It reduces household garbage disposal costs.
- It produces less odor and attracts fewer pests than putting food wastes into a garbage container.
- It saves the water and electricity that kitchen sink garbage disposal units consume.

- ☆ It produces a free, high-quality soil amendment (compost).
- ☆ It requires little space, labor, or maintenance.
- ☆ It spawns free worms for fishing.

Earthworm can be fed all forms of food waste, yard and garden waste, paper and cardboard *etc.* Leaves are dominant organic waste in most backyard compost piles. Approximately 1kg of actual nitrogen is required for 100 kg of dry sawdust.

8.6 Influencing Factors

8.6.1. Environmental Parameters

Various species of earthworms have different environmental requirements which are necessary for their propagation and continued health. These requirements will inevitably dictate whether one particular species of worms will be suitable for culture in any given circumstances. The main parameters are.

8.6.2. Air (Aeration)

Worms need to breath, just like other living creatures. The available oxygen is used up and replaced with carbon dioxide and other miscellaneous waste gases. The decreasing amount of fresh oxygen can result in an increase in heat will result in a similar rise in the oxygen requirements of worm. The whole situation is easily rectified, and only requires very infrequent attention.

8.6.3 Moisture Content

The bin content should be kept moist but no soaked. A straw covering may be needed in exposed sites to keep the bin from drying out during hot summer weather. Under natural condition, the greatest abundance of earthworms will be located in soil which average between 12 per cent -30 per cent moisture content. If the amount of available moisture should fall too low, the earthworm will begin to lose its internal water content, and a series of biological events will begin to occur, which if unchecked, will eventually result in the death of the animal. Adequate moisture is essential for microbial activity. A dry compost pile will not decompose efficiently. Enough water should be added to completely moisten the pile, but over watering should be avoided.

8.6.4 Temperature

Temperature requirement for optimal results is 20-30°C.However survival of earthworms is even at lower temperature and up to 48°C air temperature. The temperature drops below 10°C the amount of food eaten by the worms will also decrease. The worms will less active, and possibly move a little lower into the bedding causing the problem, in which case they will move nearer the surface. At the 4-50°C, the adult worms may stop producing cocoons, and the growth rate of younger worms will diminish. Red worms can survive a wide range of temperature

(40- 80°C), but they reproduce and process food waste at an optimum bedding temperature range of 20°C.

8.6.5 pH

During the initial stage of decomposition organic acids are produced, decreasing the pH. The end of composting is usually alkaline (pH 7.1 - 7.5).

8.6.7 Other Parameters

Other important parameters for potential working of earthworms on suitable waste are:

8.6.7.1. Suitable

Some parameters can be classified according to their suitability for earthworms that are:

- Kitchen waste suitable for worms includes coffee grounds and paper filters, tea bags, plate scrapings, rotting fruit (including citrus fruit but not citrus peel), vegetable peels, leftovers, moldy bread, *etc.* These materials can be raw or cooked. They do not have to be ground up, as the micro-organisms in the bin will gradually soften them.
- Chopping fruit and vegetable scraps into smaller pieces will speed the composting process. If a large quantity of dry food (*e.g.*, moldy bread) is added and covered with bedding, pour a little purified water over the bedding to moisten the mixture.
- If too much kitchen waste is added, the bin mixture putrefies before the worms can process it and becomes harmful to the worms.
- High-protein foods like beans are particularly susceptible of their mix to only include fruit and vegetable matter, avoiding grains, proteins or prepared food scraps altogether.
- Check the bin at least once a week, give the materials a stir to oxygenate, and add bedding if the bin appears too moist.
- Soft vegetable are decomposable but the worms will not process the woody parts or large roots and these will have to be hand-removed later from the finished vermicompost.
- Compost able plates, cups, *etc.* are also suitable, but in small bins they should be torn first into smaller pieces so as not to block oxygen flow.
    - **Bedding:** Bedding is the living medium and also a food source for the worms. It should be material high in carbon and made to mimic decaying dried leaves on the forest floor, the worms' natural habitat. The bedding should be moist (similar to the consistency of a wrung-out sponge) and loose to enable the worms to breathe and to facilitate aerobic decomposition of the food that is buried in it. A wide variety of bedding materials can be used, including shredded newspaper,

sawdust, hay, cardboard, coir, burlap coffee sacks, peat moss, pre-composted (aged) manure, and dried leaves. Cat litter, and pet and human waste should not be used, because they may carry disease. Most vermicomposters avoid using glossy paper from newspapers and magazines, junk mail, and shredded paper from offices, because they may contain toxins which may disrupt the system. Also, coated cardboard that contains wax or plastic, such as milk boxes, should not be used. Newspaper and phone books printed on regular, non-glossy paper with non-toxic soy ink are safe for use, and decompose relatively quickly.

- **Food:** Worms and other composting organisms have a preferred ratio of carbon to nitrogen (C : N), approximately 30:1. As some waste is richer in carbon and others in nitrogen, waste must be mixed to approximate the ideal ratio. Brown matter, or wood products such as shredded papers, is rich in carbon. Green matter, such as food scraps, has more nitrogen, which is related to the amount of protein in the waste. If the waste is mostly vegetable and fruit scraps, and does not regularly include animal products or high-protein vegetable foods like beans, the resulting vermicompost and waste liquid will be low in nitrogen.

8.6.7.2 Unsuitable

- Many materials which are not digested by the worms they are called unsuitable like Protein, fats in meat scrap these materials can attract scavengers.
- Grass clippings and other products sprayed with pesticides should be avoided.
- Too much oil or fat can hinder the breathing of the worms; meat and dairy products increases the difficulty of maintaining a healthy, low-odor vermicomposting mix, and is usually not recommended.
- Acidic foods (tomatoes, citrus), starchy foods (bread, rice), garlic and onions should only be added in moderation. Large amounts of these materials can change the balance of the system.

8.7 Applications of Vermicomposting

Vermicomposting have lots of applications, which are following:

8.7.1. Medical Application of Vermicomposting/Vermiculture Biotechnology

The medical value of earthworms dating from at least as back as 2600 B.C. and covering a range of disease from Pyorrhea to post partial weakness, from Jaundice to increase in the sperm count and excellent aphrodisiacs.

8.7.2 Earthworm and Pollution Control

The urban and rural wastes being generated continuously are undesirable pollutants for the environment and a menace to health of the community. The sources of toxic substances reaching the soil surface are mainly the solid wastes containing heavy metals released from industrial and pesticides used for health and agriculture. The accumulation of toxic chemicals in earthworm tissue is very significant ecologically because these animals are important components in the food chain of several species of birds and mammals. This is very important from the point of minimization of soil pollution. It helps for the people like job and increase the economic value.

8.7.3. Role of Vermicompost in Sustainable Agriculture

In tropical countries due to high temperature and low moisture, the use of inorganic fertilizers is not economical as there is only two fold increase in yield on application of 10 fold increased use of inorganic fertilizers. The organic carbon content is very low in tropical soil which is very important for soil organisms who contribute towards soil physical properties like aggregate stability, porosity, bulk density and water holding capacity. They also contribute towards immobilization as well as solubilization and mobilization of nutrients as and when required. Thus vermicompost which can be prepared without large scale investment can be utilized as an organic amendment to enhance biological process in soil. Thus primarily vermicompost can act as medicine for the health of soil and secondarily as a nutrient supplier to the crop.

8.8 Vermicomposting Process

- It is a process of decomposition of organic matters into humic rich manure by the action of earthworms.
- Partly degraded Cow dung, Sheep dung, Elephant dung, Pressmud, Cane trashes, City solid wastes, Poultry farm wastes and other domestic wastes are brought to Vermicomposting unit.
- There, it is formed as windrows of 1-2 feet height.
- The windrows are provided with Greenhouse shade in order to prevent from sunlight.
- Water is sprayed on the windrows by Drip irrigation or manual system to maintain the optimum moisture of 50-55 per cent.
- A special type of African Earthworm (*Udrillus ugenous*) is inoculated on the windrows.
- It starts acting on the partly decomposed materials in the windrows to give humic rich Vermicastings.
- This process takes place to complete in 60-90 days. The time period depends upon no. of worms/square meter. Simple 1gm of weight of worms collects 5gm of waste material per day.

- After completion of the process, harvesting the Earthworms from Vermicastings and earthworms are then inoculated in new windrows or top materials can be harvested layer by layer as and when vermicastings appeared.

8.9 Advantages of Vermicomposting

- Organic wastes are broken down and fragmented rapidly through earthworms resulting in stable, non-toxic material of potentially high economic value, which is used as a soil conditioner/fertilizer in agricultural and horticultural practices.
- As with the composting process, vermicomposting provides a reduction in waste bulk density.
- As an aerobic process composting leads to Nitrogen mineralization. In vermicomposting, earthworms increase and accelerate Nitrogen mineralization rate. The humification process is greater and faster during vermicomposting.
- A decrease in fulvic acid carbon and increase in the percentage of humic acid carbon are observed in vermicompost. The humic substances show an increase of 40-60 per cent, which was higher than the value obtained through composting process.
- Vermicomposting also brings down the availability of heavy metals compared to composting.
- Vermicomposting has hormone like compounds (auxins), which accelerate plant growth.
- Higher percentages of germination have been recorded with vermicompost than with compost.
- Chemical analysis of vermicomposts and composts have shown slightly lower pH value for vermicompost and slightly higher nutrient concentrations, particularly of Nitrogen. It is also quite common for vermicomposts to have very low concentration of Ammonia-nitrogen and very high concentration of Nitrate-nitrogen, whereas the opposite is true for many types of compost. The nutritional value of vermin-compost depends upon the raw materials used for feeding.

8.10 Composting of Industrial Waste and other Organic Residues

8.10.1 Phospho-Compost

The phospho-compost is a valuable amendment which could be prepared by mixing farm wastes, cattle dung, soil, compost, chopped grasses, crop residues and plant leaves with mussoorie rock phosphate at the rate of 30 per cent of the compostable materials. This mixture is made into a slurry so as to provide adequate moisture and after uniform mixing with decomposer fungus/bacteria. The slurry is allowed to decompose in a compost pit for about 60 to 90 days. The moisture is maintained at 60 per cent throughout the period of composting and the compost will be ready for use in 60 - 90 days. Once the compost is ready, 4kg/tonne of solid, liquid 400 ml/tonne of Phosphate solubizling bacteria is added and allowed for 21 days for release of fixed phosphate. Phospho-compost has better performance than other sources of phosphorus and it can be also used as an effective organic amendment in problematic soils.

8.10.2 Potassic Compost

Potassic compost is a valuable amendment to the soil which could be prepared by mixing rich farm wastes, cattle dung, sugar cane trash, soil, potassium rich ores, banana wastes *etc.* The waste materials first decomposed with the help of fungus/ bacteria using 400ml liquid cultures/tonne of waste. The use of decomposer culture decompose the materials within 45 days. About 60 per cent moisture is maintained throughout decomposition process. Once the material is decomposed, 200ml/ tonne of waste material potash mobilizing bacteria is to be added and allowed for 20 days. The compost will be rich in potassium, and hormones required for plant growth. Potassic-compost has better performance than other sources of potassium, as it is in organic form.

8.10.3 Sugar Factory Wastes and Sugarcane Trash

The major sugar factory wastes are pressmud, filter cakes bagasse and molasses. Pressmud contains about 1.2 per cent N, 3.82 per cent P_2O_2, 1.42 per cent K_2O and 11.1 per cent CaO (Palaniappan and Natarajan, 1993) and is a good source of organic matter (4.4 to 35.8 per cent). Because of high calcium content in the pressmud, it is used mainly to reclaim acidic soils substituting gypsum to some extent. There are two forms of pressmud: pressmud with carbonate process having $CaCO_3$ (66 per cent) and organic matter (4.4 per cent); presumed with sulphitation process having CaSO4 (9.3 per cent) and organic matter (35.8 per cent). The amount equivalent to one kg of 100 per cent pure gypsum will be 0.88 kg and 10.75 kg of carbonated and sulphated pressmud, respectivley and in terms of sulphur 4.0 and 57.8 kg, respectively (Singh and Dagar, 1993). Addition of pressmud improves soil aeration and drainage in heavy soils, whereas in sandy soils it helps in improving the retention of moisture. When added to sugarcane fields it increased the cane yield, improved the juice quality and enhanced the ammonifying power of the soils. In a pot experiment, pomogranate (*Punica granatum*) and *Eucalyptus tereticornis* grown in highly alkali soil (pH 10.5) when amended with pressmud could perform very well. The sugarcane trash removed during cropping to prevent lodging of the canes and the dead leaves collected at the time of cane harvest are generally burnt in the field itself. The valuable nutrients are lost during the process. The trash may be converted into a very useful organic amendment by proper composting and adding mussoorie rock phosphate at the time of composting. As the pressmud itself is rich in Potash and Phosphorus, use of 400ml/tonne of composted material with Potash Mobilizing Bacteria (*Fraturia aurientia*) and Phosphate Solubilizing

Bacteria increases the availability of K_2O and P_2O_5 in the wastes. This may be used later like other organic amendments.

8.10.4 Oil Cakes

Many oil cakes such as the castor, neem, madhuca, karanja, linseed, rape seed and cotton seed which are non-edible oil cakes may serve as useful organic manure as these contain high amounts of plant nutrients. Most of the non-edible oil cakes are valued much for their alkaloid contents which inhibit the nitrification process in soils. Neem cake contains the alkaloids-nimbin and nimbicidine which effectively inhibit the nitrification process. Similarly, Karanjin (*Pongamia pinnata*) and (*Madhuca butyracea*) is a potent nitrification inhibitor equal in efficiency to nitropyrin inretarding the nitrification process of ammoniacal nitrogen and increasing the yield, nitrogen uptake and grain protein content of rice. Madhuca cake has been successfully used in coastal saline soils for cultivatiion of rice. soils. Neem cake contains the alkaloids-nimbin and nimbicidine which effectively inhibit the nitrification process. Similarly, Karanjin (*Pongamia pinnata*) and madhuca (*Madhuca butyracea*) is a potent nitrification inhibitor equal in efficiency to nitropyrin inretarding the nitrification process of ammonical nitrogen and increasing the yield, nitrogen uptake and grain protein content of rice.

8.10.5 Fly Ash

Fly ash is a useful organic amendment. At many places acidic mined lands exhibit reduced biological activity. In some strip mine sites with pH ranging from 2.6 to 4.7, the pH was found to range between 5.0 and 7.5 after 10 years of treatment with fly ash and the credit of vegetation on these sites was rated fair to good. With the exception of N, fly ash is rich in P_2O_5, K_2O, Ca and Mg and also contains elements like B, Mo and Zn which are essential for plant growth and are available to the vegetation raised on soils treated with fly ash. Mixing of potash mobilizing bacteria and phosphorus solubilizing bacteria at the rate of 400ml/tonne increase the availability of P_2O_5 and K_2O within 20 days

8.10.6 Industrial Effluents

Effluents from sugar factory, fertilizer factory and paper industry are rich in plant nutrients and can be used after proper treatments. The sewage sludge has been found to be as effective in increasing the yield of crops as inorganic fertilizers and also of forest tree species such as Eucalyptus and Populous. But from the sewage and industrial effluents toxic substances and heavy metals are to be removed completely as otherwise they can cause health-problems when they reach in food chain. Use of proper decomposing microorganisms also reduces the heavy metals. After decomposition mixing of potash mobilizing bacteria and phosphorus solubilizing bacteria at the rate of 400ml/tonne increase the availability of P_2O_5 and K_2O with in 20 days.

8.10.7 Coir Pith Compost

The coir dust is a waste product of the coir industry and could be used as organic amendment. the coir waste accumulates in large quantities near the coir industrial units and it can be converted into valuable organic manure by proper composting with the aid of the mushroom fungus Pleurotus sojorcaju. The composting reduces the volume by 42 per cent besides narrowing the C: N ratio. Once the compost is matured add potash mobilizing bacteria and phosphorus solubilizing bacteria at the rate of 400ml/tonne increase the availability of P_2O_5 and K_2O with in 20 days.It has been found to be a good amendment for sodic soils and a moisture conserving material for rainfed crops. The use of coir pith without composting is, however, not advisable. In the fresh form it contains 8 - 12 per cent soluble tannin - related phenolics which can inhibit microbial activity in the soil and cause immobilisation of nitrogen. Table 8.2 shows Composition of raw and composed coir pith.

Table 8.2. Composition of Raw and Composed Coir Pith

Properties	*Raw Coirpith*	*Composed coirpith*
Lignin (per cent)	30.00	4.80
Cellulose (per cent)	26.52	10.10
Organic carbon (per cent)	29.00	24.90
N (per cent)	0.26	1.06
P2O5 (per cent)	0.01	0.06
K2O (per cent)	0.78	1.20
Ca (per cent)	0.40	0.50
Mg (per cent)	0.36	0.48
Fe (per cent)	0.07	0.09
Mn (ppm)	12.50	25.00
Zn (ppm)	7.50	15.80
Cu (ppm)	3.10	6.20
C:N ratio	112:1	24:1

8.10.8 Poultry Waste

Compost Poultry waste comprises waste feed, solid and liquid dropping, litter, eggshell, diseased and dead birds, culled birds, feathers and the wastes from poultry sheds. Poultry waste management is highly complex and challenging because of associated problems like nitrate and heavy metal contamination in soil, crops, surface and ground water, air quality and odor; disposal of dead and diseased poultry and food safety. Poultry manure is good source of nutrients, particularly manure is good source of nutrients, particularly for vegetable. The decomposition of the waste must be done with the use of fungus/bacteria. Once the decomposition is completed, KMB, PSM and Paceli-T.V. in liquid 100ml/tonne suppresses the disease causing pathogens. Table 8.3 shows nutrient status in poultry waste.

Table 8.3. Plant Nutrient Status in Poultry Waste

Source	*Per cent*			
	Nitrogen	*Ammonia*	*Phosphorus*	*Potassium*
Fresh chicken manure	3.7-8.8	0.4-1.1	1.2-2.0	1.2-2.7
Poultry litter manure	1.4-6.8	0.5-1.1	0.5-3.51	1-2.7
Broiler litter[b]	2.3-6.0	-	0.6-3.9	0.7-5.2

a. Adapted from Sims and Wolf (1994); b. Broiler litter also contains 23-125 boron, 25-1003 copper, 125- 667 manganese and 106-669 zinc in parts per million.

8.10.9 Saw Dust

It is possible to use waste products from timber industry as organic matter. Inoculation with cellulolytic strains of *Bacillus* sp., *Cephalosporium* sp. and *Streptomyces* sp. has been reported to increase the decomposition of saw dust and bark. Mixing of potash mobilizing bacteria and phosphorus solubilizing bacteria at the rate of 400ml/tonne increase the availability of P_2O_5 and K_2O with in 20 days.

The following criteria for judging a good quality compost may be as follows:

- ☆ Dark brown to black.
- ☆ Crumbly structure not packed or lumpy.
- ☆ Largely insoluble in water.
- ☆ C:N ratio from 10 to 20.
- ☆ Dissolves to a large extent in dilute alkali.
- ☆ Beneficial effect upon soil and growing vegetation.
- ☆ Pleasant earthy smell.
- ☆ Will not attract flies.

8.11 Biogas Slurry Applications

Biogas slurry is good source of organic manure. The following are the different methods of applying biodigested slurry as manure:

- ☆ Air dried biogas slurry can be applied by spreading on the agricultural land at least one week before sowing the seeds or transplanting the seedlings.
- ☆ The liquid slurry can be mixed directly with the running water in irrigation canal which will enable spreading of the slurry uniformly in the cropped area or in cultivation land
- ☆ Biogas slurry can also be coated on the seeds prior to sowing. This acts as insecticide and prevents seeds or plants from insect attack. This helps in early germination and healthy growth of seedlings.

- ☆ The digested slurry is fed through the channel, flowing over a layer of green or dry leaves and filtered in the bed. The water from the slurry filters down which can be reused for preparing another fresh dung slurry. The semi-solid slurry can be transported easily as it was in the consistency of fresh dung and used for top dressing of crops like sugarcane and potato.
- ☆ Biodigested slurry is also being used to fish culture, which acts as a supplementary feed. On an average, 15-25 litre of wet slurry can be applied per day in a 1200 sq pond. Slurry mixed with oil cake or rice-bran in the 2-1 ratio increases the fish production remarkably. In general, organic manures about 10t/ha in the form of FYM or compost or biodigested slurry is recommended to be applied once in three years to maintain the organic content of soil, besides providing nitrogen, phosphorus and potassium in the form of organic fertilizers to the crop.
- ☆ The digested slurry, if mixed with Azospirillum, KMB and PSM @ 200ml each/acre ensures increase of yield minimum 30 per cent over slurry alone. Table 8.4 Nutrient content in biogas slurry and FYM.

Table 8.4. Nutrient Content in Biogas Slurry and FYM

Particulars	*Nutrient Content*						
	Major Element (per cent)				*Trace Element (ppm)*		
	N	*P*	*K*	*Fe*	*Mn*	*Zn*	*Cu*
Biogas slurry	1.43	1.21	1.01	4200	550	150	52
Farm Yard manure	0.94	0.56	0.72	4000	490	100	45

Leaf Compost

Leaf composting, can be achieved by heap or ditch composting or by windrow composting. Windrow are preferred as they allow efficient handling of materials. Provide good aeration, allow sufficient absorption of water and are easy to be formed.

Formation of uniform-shaped windrow from 2.40m – 3.60 m at the base and 2.40m-3.00m high and of any convenient length. Windrow built too high will have excessive compaction at the base resulting in anaerobic conditions. Windrows built too low will not allow sufficient insulation to sustain thermophilic temperature during cold weather. To ensure proper aeration, it is important to break apart tightly compacted leaves.

Though reduction in size may aid in rapid decomposition, it is not desirable in leaves because it increases the compaction making more frequent aeration necessary. When the incoming leaves are not adequately moist, it is desirable to add water to maintain a proper moisture regime of 40 to 60 per cent.

The C:N ratio of leaves is relatively high. It can be as high as 80, and needs to be amended with nitrogen. Sewage sludge, green manure and grass clippings are good sources of nitrogen. Proper aeration can be maintained by periodical mixing of the material.

Under optimum environmental composting conditions, leaf compost will be ready between 180 and 270 days. Leaf mould will have pH range of 6-7.

Further the use of finished compost (leaf mould) as covering material (10-15 cm) in the subsequent preparation of leaf compost to supply a heavy inoculum of micro-organisms.

8.12 Use of Biofertilisers in Composting

Biofertilisers are microbial preparations which on application to soil help in augmenting agricultural production. Preparations containing cellulose decomposing, phosphate solubilizing, potash mobilizing and nitrogen fixing microorganisms come under biofertiliser. Azotobacter, a nitrogen fixer, and phosphate solubilising microorganisim, Fraturia ausintia, a potash mobilizing microorganism, Aspergillus awamori and Bacillus polymyxa are heterotrophs. Inoculation of compost, one month after starting the process, when thermophilic activities have subsided at around 30 - 38°C, with Azotobacter and phosphate solubilising and potash mobilizing bacteria biofertilisers in rice straw + rock phosphate (3 per cent of substrate) improves the nitrogen content and phosphorus and potash availability of the mature compost.

8.13 Biodynamic Preparations

The (biodynamic - energy) means working with the energies which create and maintain the life. Biodynamic is not bio-organic because biodynamic preparation do not contain plants nutrients and are required in every small quantities after energising and many fold dilutions.

There are eight known biodynamic preparations, namely:

1. Biodynamic preparation 500 (Cow-horn manure),
2. 501 (Horn silica),
3. 502 (Yarrow),
4. 503 (Chamobile),
5. 504 (Stinging nettle),
6. 505 (Oak bark),
7. 506 (Dandelion), and
8. 507 (Valerian).

These preparations are easy to formulate and can be done by farmers at their own farms.

To Increase Humus in Soil (Cow-horn manure)

Ingredients	*Amount*	*Methods*	*Results/Effects*
1. Dung from lactating cows; 2. Horns of the cows	Sufficient cow-dung to pack the cow horns (about 500gms in one horn) in Autumn (Sept.-Oct)	Six months fermentation in the soil (during ascending moon days in autumn (Sept.–Oct) leaving to accumulation of fecal bacteria and humus forming bacteria. The property of this dung is similar to that of earth-worm castings. It should be dug out during descending moon days in Spring *i.e.*, April-May.	Increase in NO_3-N contents (from 0.06 per cent to 1.7 per cent) in the soil. Promotion of humus forming process and better root growth. An increase of 75 per cent in O_2 absorption. Bacterial count 500million/gm of farm soil or pasture. More active root system resulting in promotion in vegetative growth.

Applications

After dilution in water, it is sprayed directly on land during autumn(March,April-May and early spring September-early October) concentration 30-35 gms in 12 litres of boiled cool water and stirred for one hour in the evening before sowing or transplanting.Mixing of PSM, KMB, Azotobacter, Azosprillium and Bacillus subtilis 100 ml each in solution ensures better yields in all crops.

To Increase Photosynthesis (Horn silica)

Ingredients	*Amount*	*Methods*	*Results/Effects*
Finely powdered silica (quartz)	Silica powder in one horn 500-600g	Treated for 6 months in cow horns in the soil and used as foliar spray. The horn is buried on full moon day or descending moon.	Growth of stem and leaf as well as assimilation process stimulated. Increase in assimilation effect in leaves up to 3.5 times. In the powdered rock silica aerobic bacteria count increases from 0 to 40 million per gram. Photosynthesis enhanced, resulting in good yield. Improvement In nutrient quality of fruits and grains. After spray the plants response to atmospheric environmental factors like air, light, temperature *etc.* increases.

Applications

After dilution in water, it is sprayed directly on acres land during early spring October,concentration 30-35 gms in 12 litres of boiled cool water and stirred for one hour in the evening before sowing or transplanting. Mixing of PSM,KMB, Azotobacter, Azosprillium and *Bacillus subtilis* 100 ml each in solution ensures better yields in all leafy crops like tea, cabbage, spinach *etc.*

To Increase Metabolic Activity in Soil (Preparation 502)

Ingredients	*Amount*	*Methods*	*Results/Effects*
1. Yarrow blossoms (*Archillea milefolium*) 2. Male Deer (stag) bladder	1 kg blossoms in 1 bladder	The blossoms are fermented together with deer bladder for 6 months in soil during winter. Before that, it is hanged from a tree for 5-6 months for absorbing energy from the sky.	Use of sulphur potassium by the plant (for their growth) stimulated resulting building up of proteins and carbohydrates is also a 'bio-catalyst. This has 9-10 millions aerobic bacteria.

Applications

After dilution in water, it is sprayed directly in one acre land during concentration 30-35 gms in 12 litres of boiled cool water and stirred for one hour in the evening before sowing or transplanting. Mixing of PSM, KMB, Azotobacter, Azosprillium and Bacillus subtilis 100 ml each in solution ensures better yields in all crops.

Soil Conditions (Preparation 503)

Ingredients	*Amount*	*Methods*	*Results/Effects*
1. Chamomile blossoms (*Matricaria recutita*) 2. Small intestines of cows 3. Good soil humus (for enriching soil building bacteria)	Blossoms in one intestine 500-600gms	Both are fermented together inside the soil in winter	Increase in aerobic bacterial count to the extent of million no anaerobic growth. Increase in microbial population in the soil improves soil physical and chemical conditions. The net results may be observed after continuous use of preparation up-to 4 years.

Applications

After dilution in water, it is sprayed directly in one acre land during concentration 40-50 gms in 12 litres of boiled cool water and stirred for one hour in the evening before sowing or transplanting.Mixing of PSM,KMB, Azotobacter,Azosprillium and Bacillus subtilis 100 ml each in solution ensures better yields in all crops

To Increase N-Fixing Bacteria Activity (Preparation 504)

Ingredients	*Amount*	*Methods*	*Results/Effects*
1. Stinging nettle (*Urtica dioica*) leaves and stems	In one place 1-2 kg	Buried about 2 feet deep in soil humus separated from soil by a thin layer of peat moss	Stinging nettle is rich in vitamins and iron. Its humus is enriched (about100 times) molybdenum and vanadium which are trace elementary necessary for the activity of nitrogen fixing bacteria. The bacterial count around Fifty millionone billion aerobic 470 million anaerobic per gram.

Applications

After dilution in water, it is sprayed directly in one acre concentration 50 gms in 12 litres of boiled cool water and stirred for one hour in the evening before sowing or transplanting. Mixing of PSM,KMB, Azotobacter, Azosprillium and Bacillus subtilis 100 ml each in solution ensures better yields in all crops.

Disease Resistance and NPK-Enricher to Soil (Preparation 505)

Ingredients	*Amount*	*Methods*	*Results/Effects*
1. Oak bark (Quercus robur or Quercus alba from not too old tree)	500-600g in one skull	Buried together resulting in dark blackish brown humus of a fine structure	Highest calcium ash content in oak trees growing even in sandy Ca deficient soil. Ca content is very high (more than 10 per cent) of which 0.1 per cent is readily available. K content 0.13 per cent, PO_4 content 0.03 per cent in available fraction.
2. Skull of cows			Bacterial count two billion aerobic and 70 million anaerobic. It develops disease resistance in plants.

Applications

After dilution in water, it is sprayed directly in one acre land during concentration 50 gms in 12 litres of boiled cool water and stirred for one hour in the evening before sowing or transplanting. Mixing of PSM,KMB, Azotobacter, Azosprillium and *Bacillus subtilis* 100 ml each in solution ensures better yields in all crops.

Preparation 506

Ingredients	*Amount*	*Methods*	*Results/Effects*
Dandelion (*Taraxacum officianale*)	500-600gm blossoms in one mesentery	Pack the blossoms in mesentery, tie it and bury it soil below 2' deep.	This has Ca, Mg, NO_3.N, PO_4 and SO_4 Bacterial count 360 million (aerobic) + a spreading species and 180 million anaerobic/ gram of soil.

Applications

After dilution in water, it is sprayed directly in one acre land during concentration 50 gms in 12 litres of boiled cool water and stirred for one hour in the evening before sowing or transplanting. Mixing of PSM, KMB, Azotobacter, Azosprillium and *Bacillus subtilis* 100 ml each in solution ensures better yields in all crops.

Compost Enricher (Preparation 507)

Ingredients	*Amount*	*Methods*	*Results/Effects*
Pressed out cold extract from Valeriana flowers which is used for spastic nervous condition	1-2 kg Valeriana blossoms	In high dilution (5-10 drops/10 lit. of water) as spray over the base and cover of manure and compost heaps	Attracts earthworms and stimulates their propagation. This has one million aerobic bacteria. *Juice Composition* K-0.335 per cent; Ca-0.425 per cent; Mg-0.005 per cent; NO_3-N 0.145 per cent; NH_4.N showed an increase from 0.02 per cent to 0.6 per cent; PO_4-0.06 per cent; Mn-Trace

Applications

After dilution in water, it is sprayed directly in one acres land during Winter,concentration 30-35 gms in 12 litres of boiled cool water and stirred for one hour in the evening before sowing or transplanting. Mixing of PSM,KMB,

Azotobacter, Azosprillium and Bacillus subtilis 100 ml each in solution ensures better yields in all crops.

Preparation 508

Ingredients	*Amount*	*Methods*	*Results/Effects*
Dried herb of *Equisetum arvenae* (Horse tail)	2-3 kg herb	Mix it in 100 litres of water and keep for 1-2 months and then spray	Mineral contents Silicon 10 per cent; K 1.15 per cent; Ca 0.42 per cent; Mg 0.01 per cent; NO_3-N 0.47 per cent; NH_3–N Trace PO_4 0.06 per cent; SO_4 0.36 per cent; Aerobic bacterial count; 21000/g; No anaerobic bacteia

Applications

After dilution in water, it is sprayed directly in one acre land during Winter,concentration 30-35 gms in 12 litres of boiled cool water and stirred for one hour in the evening before sowing or transplanting. Mixing of PSM,KMB, Azotobacter,Azosprillium and Bacillus subtilis 100 ml each in solution ensures better yields in all crops.

Biodynamic with Carriers

Different carriers can be used for commercialization and preservation of different biodynamics for longer shelf life. It was observed that, different products of biodynamic are available in the market of our own country as well as some imported products. The carrier used talcum powder, Kelvin powder, china clay *etc.* as a carrier after mixing 15-20 per cent of the preparation, which can be kept for minimum 6 months. Otherwise, the preparation have to be used quickly to avoid deterioration of the product. For the preservation of the fresh product, the container has to be kept in cool place away from direct sunlight at 200 c temperature. Some products are available in the market on the name of Energy, bacterial preparation, *etc.* It is nothing but only biodynamic preparation.

Table 8.5. Status of Organic Manures and Soil Conditioners in Organic Farming

Sl.No.	*Substances Description, Compositional Requirements*	*Conditions for Use*
I	**Plant and Animal Origin**	
1	Farmyard manure, slurry and urine	Permitted
2	Vermicastings/Vermicmpost	Permitted
3	Blood meal, meat meal, bone, bone meal	Permitted
4	Hoof, and horn meal, feather meal, fish and fish products, wool, fur, hair, dairy products	Permitted

Sl.No.	Substances Description, Compositional Requirements	Conditions for Use
5	Biogradable processing by products, plant or animal origin *e.g.* by product of food, feed, oilseed, brewery, distillery or textile processing	Permitted
6	Crop and vegetable residues, mulch, green manure, straw	Permitted
7	Wood, bark, sawdust, wood shavings, woo dash, wood charcoal	Permitted
8	Seaweed and seaweed products	Permitted
9	Plant preparations extracts	Permitted
10	Urban composts from separate sources which are monitored for contamination	Permitted
II	**Mineral Origin**	
1	Basic slag	Permitted
2	Soil amendments like lime, limestone, gypsum, Calcarious and magnesium soil amendments.	Permitted
3	Magnesium rock, kieserite and Epson salt (magnesium sulphate)	Permitted
4	Naturally occurring potassium minerals	Shall be obtained by physical procedures but not enriched by chemical process.
5	Natural phosphates (like rock phosphate)	Permitted
6	Trace element	Permitted
7	Sulphur	Permitted
III	**Microbiological**	
1	Biodegradable processing by-products of microbial origin, *e.g.* by-products of brewery or distillery processing, microbiological preparations based on naturally occurring organisms – Biofertilizers	Permitted
IV	**Others**	
1	Biodynamic preparations	Permitted

8.14 Standards for Inputs to be Used in Organic Agriculture

Organic Agriculture uses a lot of inputs, which are basically organic in nature and are expected to be produced on the farm where the farming is being done. There is a wide variation in the quality of inputs that are currently used in organic farms. It is therefore important and necessary that efforts should be taken up develop standards for these inputs. An initial effort is being made here to do the same.

Compost: "A solid mature product resulting from composting, which is managed process of biooxidation of a solid heterogeneous organic substrate including a thermophilic phase."

There are three criteria applicable as standards as standards for correct compost:

8.14.1 Maturity

Several indicators are available to determine compost maturity. Presently the indicators to be applied are:

1. Compost must be cured for at least 15 days.
2. Compost will not reheat after the composting process to greater than 20^{0}C above ambient temperature.

8.14.2 Trace Elements

Trace elements are defined as a chemical element present in compost at a very low concentration. The compost standards identify trace elements that are essential to plant growth in addition to identifying heavy metals which depending on their concentration in the soil could be harmful to human health and the environment.

8.14.3 Pathogens

Pathogenic organisms are sometimes present in the feed stocks used make compost. As result the compost may also contain pathogens. To reduce any potential health concerns, treatment processes as well as biological specifications have been identified. All the pathogens generally present in the compost feed stock get de-activated at a temperature of around 55°C. Hence it is recommended that:

- Using universal composting method, the solid waste shall be maintained at operating conditions at 55°C or higher for three days.
- Using WINDROW composting method, the solid waste attain a temperature of 55°C or higher for atleast 15 days during the composting period. Also during the high temperature period the WINDROW shall be turned at least five times.
- Using an aerated static pile composting method, the solid waste shall be maintained at operating conditions of 55°C or higher temperature for three days. The preferable practice is to cover the pile with an insulating layer of material such as cured compost or wood chips to ensure that all areas of feed material are exposed to the required temperature.

While organic farming envisages the production of required inputs in the farm itself, there will be instances particularly during in-conversion period when organic inputs from other farms are obtained. If these inputs are obtained from other organic farms then the above standards will apply.

Chapter 9

Integrated Disease and Pest Management

9.1 Introduction: An Overview

Integrated pest management, or IPM, is a process you can use to solve pest problems while minimizing risks to people and the environment. IPM can be used to manage all kinds of pests' anywhere–in urban, agricultural, and wild land or natural areas.

Integrated Pest Management (IPM) is an effective and environmentally sensitive approach to pest management that relies on a combination of common-sense practices. IPM programs use current, comprehensive information on the life cycles of pests and their interaction with the environment. This information, in combination with available pest control methods, is used to manage pest damage

Biological control of pest is the non-chemical way of killing pest which is environmentally friendly. In recent years, microbial antagonists are used for the control of pest and diseases. Typical example abounds on the Bacillus thuringiensis which are toxic to many species of insects. In addition, Entomopathogenic nematodes in the families' Steinernematidae and Heterorhabditidae have been used to suppress populations of pest insects in a variety of agro ecosystems, and in several cases their positive effects on crop yield have been shown. Lastly, the advantages of use of microbial control agents include safety for humans and other no target organisms, reduction of pesticide residues in food, preservation of other natural enemies, and increased biodiversity in managed ecosystems.

by the most economical means, and with the least possible hazard to people, property, and the environment.

The IPM approach can be applied to both agricultural and non-agricultural settings, such as the home, garden, and workplace. IPM takes advantage of all appropriate pest management options including, but not limited to, the judicious use of pesticides. In contrast, organic food production applies many of the same concepts as IPM but limits the use of pesticides to those that are produced from natural sources, as opposed to synthetic chemicals.

Pest

Pests are organisms that damage or interfere with desirable plants in our fields and orchards, landscapes, or wild lands, or damage homes or other structures. Pests also include organisms that impact human or animal health. Pests may transmit disease or may be just a nuisance. A pest can be a plant (weed), vertebrate (bird, rodent, or other mammal), invertebrate (insect, tick, mite, or snail), nematode, pathogen (bacteria, virus, or fungus) that causes disease, or other unwanted organism that may harm water quality, animal life, or other parts of the ecosystem.

9.2 History of Integrated Pest Management

Michelbacher and Bacon (1952) coined the term "integrated control"

Stern *et al.* (1959) defined integrated control as "applied pest control which combines and integrates biological and chemical control"

Geier (1966) coined the term "pest management"

Council on Environmental Quality (CEQ, 1972) gave the term "Integrated Pest Management"

In 1989, IPM Task Force was established and in 1990. IPM Working Group (IPMWG) was constituted to strengthen implementation of IPM at international level.

In 1997, Smith and Adkisson were awarded the World Food Prize for pioneering work on implementation of IPM.

9.3 Definitions

IPM is an ecosystem-based strategy that focuses on long-term prevention of pests or their damage through a combination of techniques such as biological control, habitat manipulation, modification of cultural practices, and use of resistant varieties. Pesticides are used only after monitoring indicates they are needed according to established guidelines, and treatments are made with the goal of removing only the target organism. Pest control materials are selected and applied in a manner that minimizes risks to human health, beneficial and nontarget organisms, and the environment.

9.3.1 Definition by Food and Agricultural Organization (FAO, 1967)

Integrated Pest Management (IPM) is a pest management system that, in the context of associated environment and population dynamics of the pest species, utilizes all suitable techniques and methods in as compatible a manner as possible and maintains pest populations at levels below those causing economic injury.

In IPM three aspects are emphasized, *i.e.*

1. Multiple control tactics used in a compatible manner
2. The populations maintained below levels that cause economic damage
3. Socio economic and eco-friendly to the environment

9.3.2 IPM Definition by Luckmann and Metcalf (1994)

IPM is defined as the intelligent selection and use of pest control tactics that will ensure favourable economical, ecological and sociological consequences.

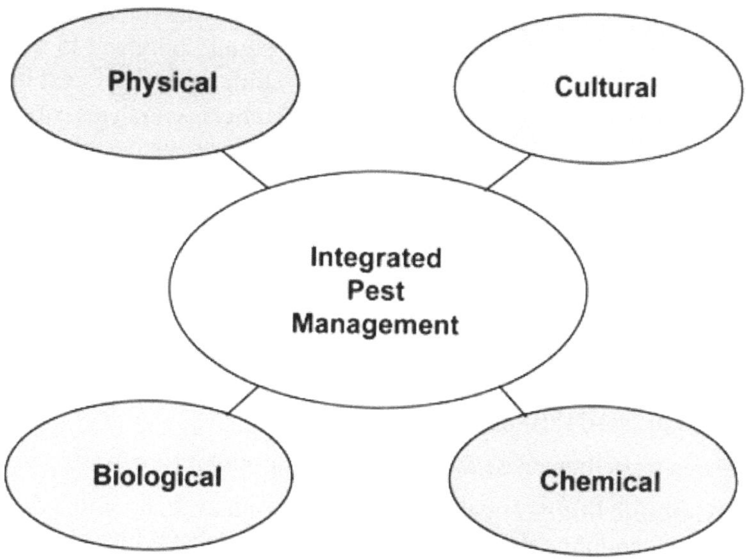

9.4 Concepts of IPM

IPM seeks to minimize the disadvantages associated with use of pesticides and maximizing socio, economic and ecological advantages.

9.4.1. Understanding the Agricultural Ecosystem

An agro-ecosystem contains a lesser diversity of animal and plant species than natural ecosystem like forests. A typical an agro-ecosystem contain only 1-4 major crop species and 6-10 major pest species. An agro-ecosystem is intensively manipulated by man and subjected to sudden alterations such as ploughing, inter cultivation and treatment with pesticides. These practices are critical in pest management as pest populations are greatly influenced by these practices. Agro ecosystem can be more susceptible to pest damage and catastrophic outbreaks owing to lack of diversity in species of plants and insects and sudden alternations imposed by weather and man. However, agro-ecosystem is a complex of food chains and food webs that interact together to produce a stable unit.

9.4.2. Planning of Agricultural Ecosystem

In IPM programme the agricultural system can be planned in terms of anticipating pest problem and also the ways to reduce them that is to integrate crop protection with crop production system. Growing of susceptible varieties should be avoided and related crops shouldn't be grown. Bhendi followed by cotton increases incidence of the spotted borer. Groudnut followed by soybean increases incidence of the leaf miner.

9.4.3. Cost Benefit Ratio

Based on the possibility of pest damage by predicting the pest problem and by defining economic threshold level, emphasis should be given to cost benefit ratio. The crop life table to provide solid information analysis of pest damage as well as cost benefit ratio in pest management. Benefit risk analysis comes when a chemical pesticide is applied in an agro-ecosystem for considering its impact on society as well as environment relevant to its benefits.

9.4.4. Tolerance of Pest Damage

The pest free crop is neither necessary in most cases for high yields nor appropriate for insect pest management. Castor crop can tolerate up-to 25 per cent defoliation. Exceptions occur in case of plant disease transmission by vectors. The relationship between density of pest population and profitability of control measures is expressed through threshold values.

The terms used to express the levels of pest population are:

a) ***Economic Injury Level (EIL):*** Lowest population at which the pest will cause economic damage or it is the pest level at which the damage can no longer be tolerated and therefore it is the level at or before which the control measures are initiated. The amount of injury which will justify the artificial control measures is termed as economic damage. EIL is usua lly expressed as the number of insects per unit area

b) ***Economic Threshold Level (ETL):*** It is the index for making pest management decisions. ETL is defined as the population density at which control measures should be applied to prevent increasing pest population from reaching the economic injury level. Relationship between EIL and ETL can be expressed as when no action is taken at ETL the population reaches or exceeds EIL. *e.g.:* ETL value for BPH in rice is 25 insects/hill; Grasshoppers or cutworms is 1 insect/hill; rice stem borer -5 per cent dead hearts; Gall midge of rice-5 per cent silver shoots.

c) ***General equilibrium position (GEP):*** It is the average population density of insect over a long period of time unaffected by temporary interventions of pest control. However the economic injury level may be at any level well above or below the general equilibrium.

9.4.5. Leaving a Pest Residue

Natural enemy population is gradually eliminated not only in the absence of their respective insect hosts because of the indiscriminate use of broad spectrum insecticides, which in turn also eliminate natural enemies. Therefore, it is an important concept of pest management, to leave a permanent pest residue below economic threshold level, so that natural enemies will survive.

9.4.6. Timing of Treatments

Treatment in terms of pesticide spray should be need based, with minimum number of sprays, timely scheduled, combined with improved techniques of pest monitoring and crop development. *e.g.*: Use of pheromone traps for monitoring of pest population.

9.4.7. Public Understanding and Acceptance

In order to deal with various pest problems special effort should be made for effective communication to the people for better understanding and acceptance of pest management practices. The IPM practices followed should be economical and sustainable.

9.5 Principles of Integrated Pest Management

IPM is a decision-making process that helps to prevent pest problems. With IPM programs, all information and treatment methods are considered in order to manage pests. This should be effective, affordable, and safe for the environment. Elements of any IPM program include:

- ***Prevention:*** Organisms are kept from becoming problems by planning and managing ecosystems.
- ***Identification:*** Pests and beneficial organisms are identified.
- ***Monitoring:*** Pest and beneficial organism's populations are watched, as well as pest damage, and the environment.
- ***Injury and Action Decision:*** Injury and action thresholds are used to know when to treat pests.
- ***Treatments:*** Treatments (or a combination) are used, including cultural, biological, physical, mechanical, behavioural, or chemical methods. The goal is to control pests with little impact on the environment.
- ***Evaluation:*** The effectiveness of pest management plans is considered.

9.6 Advantages of IPM

Advantages of IPM include:

- Long-term answers to pest problems
- Protecting environmental and human health by reducing pesticide use
- Reducing harm to beneficial organisms
- Preventing creation of pesticide resistant pests
- Providing a way to manage pests when pesticides cannot be used

9.7 Limitations of IPM

An IPM program requires a higher degree of management:

Making the decision not to use pesticides on a routine or regular basis requires advanced planning and therefore a higher degree of management. This planning includes attention to field histories to anticipate what the pest problems might be, selecting crop varieties which are resistant or tolerant to pest damage, choosing tillage systems that will suppress anticipated pest damage while giving the crop the greatest yield potential.

IPM can be more labour intensive, consistent, timely and accurate field scouting takes time. Without this information,intelligent management decision cannot make.

Success of IPM programmes can be weather dependant. Therefore good IPM planners will have a alternate plan for when these problems arise.

9.8 Economics of Integrated Pest Management

An Integrated Pest Management (IPM) programme can be successful only if reduces cost on control measures, or increases crop yield (or both) and also reduce environmental pollution and health hazards.

9.9 Biological Control in Integrated Pest Management

9.9.1 Introduction

Biological control is the beneficial action of predators, parasites, pathogens, and competitors in controlling pests and their damage. Biocontrol provided by

In nature, organism populations suffer frequent attacks and high mortality rates from predators, parasites, parasitoids, and diseases, collectively called "natural enemies." Biological control tactics use natural enemies or agents (some practitioners call them "beneficials") to manage pests. The ultimate goal of biological control is to suppress pest population and damage without pesticide or with reduced pesticide use. Natural enemies are utilized differently depending on the target pest, host, environmental condition, and pest life cycle. There are three general approaches to biological control.

these living organisms (collectively called natural enemies) is especially important for reducing the numbers of pest insects and mites. Biological control is an environmentally sound and effective means of reducing or mitigating pests and pest effects through the use of natural enemies. According to S. H. Dreistadt, 2007, biological control efforts against plants and insects have different histories, with insect biological control being used for much of its first century largely against crop pests. Only in the1990s did insect biological control against environmental pests develop as an independent goal (Van Driesche, 2008).

However, biological control has been actively practiced for more than 100 years and the history of biocontrol, its failures and successes, has been extensively reviewed. Interest in biological control has increased over recent decades for many reasons. First, a greater appreciation for environmental stewardship among regulators, growers, and the public has promoted development of more sustainable farming practices. Second, a number of arthropod pests have developed resistance to one or more pesticides leaving growers to search for alternative management strategies. Finally, consumers increasingly demand products that are grown in a sustainable manner and are free of insecticide residue. Despite this, growers have been slow to adopt biological control as part of their pest management program.

9.9.2 Strategies of Biological Control in IPM

There are 3 basics strategies in biological control of pests, *viz.*, Classical Biological Control (Importation), Augmentation and Conservation.

9.9.2.1 Classical Biological Control (Importation)

Classical biological control is the importation of pest natural enemies from other countries, to a new locale where they do not occur naturally. It is the international introduction of an exotic, usually co-evolved, biological control agent for permanent establishment and long term pest control (Pickerel 2004). The goal of classical biological control is to find useful natural enemies, introduce them into the area of the target pest, and permanently establish them so that they will provide continuing pest control with little or no additional human intervention. The search for natural enemies in other countries is often referred to as foreign exploration.

The process of importation involves:

- Determining the origin of the introduced pest.
- Collecting appropriate natural enemies associated with the pest or closely related species.
- Selected natural enemies are then passed through a rigorous assessment, testing and quarantine process, to ensure that they will work and that no unwanted organisms (such as hyper parasitoids) are introduced.
- Mass production and release of selected natural enemies.
- Follow-up studies are conducted to determine if the natural enemy becomes successfully established at the site of release, and to assess the long-term benefit of its presence.

In recent years, classical biological control has come under increasing scrutiny for its non-target effects. However, there are many examples of successful biological control, and the need for biological control is increasing. Lastly, Classical biological control is a powerful tool for suppression of invasive plants and insects in natural ecosystems. It will play an increasingly important part in ecological restoration because; it provides a means to permanently suppress invaders over large landscapes without long-term resource commitments and hence is sustainable (M.S. Hoddle, 2003). As such, it merits use against many invasive plants and insects that are environmental pests in sensitive landscapes.

9.9.2.2 Augmentation Biological Control

Augmentation is the periodic release of a natural enemy that does not occur naturally in sufficient numbers to keep a pest below damaging levels. It's also defined as the release of additional numbers of a natural enemy when too few are present to control a pest effectively. The practice of augmentation is based on the knowledge or assumption that in some situations there are not adequate numbers or species of natural enemies to provide optimal biological control, but that the numbers can be increased (and control improved) by releases. This relies on an ability to mass-produce large numbers of the natural enemy in a laboratory or by companies to produce and sell them. There are two general approaches to augmentation: inundative releases and inoculative releases.

Inundative Releases

Inundation involves releasing large numbers of natural enemies for immediate reduction of a damaging or near damaging pest population. It is a corrective measure; the expected outcome is immediate pest control. The inundative approach is achieved by flooding the crop with multiple releases of insectary-reared natural enemies. The released insects control pests present at the time, but there is little expectation that later generations will persist at sufficient levels to provide control. In practice, releases are often repeated if pest populations were not all present in a susceptible stage during the previous application, if new pests disperse into the crop, or if the crop is long lived, increasing the length of time it could become infested. Additionally, inundative release of natural enemies is undertaken, causing effects similar to the use of conventional insecticides, as there is a knockdown effect of the target host population. Therefore, it may be used in the field and in greenhouse as seasonal release. However, because of the nature of natural enemy activity, and the cost of purchasing them, this approach using predaceous and parasitic insects is recommended only in certain situations, such as the mass release of the egg parasite Trichogramma for controlling the eggs of various types of moths. *Trichogramma evenescen* Westwood, *T. brassicae* Bezdenko, T. cacoeciae Marchal, and *T. dendrolimi* Matsumura are sold in Europe to control Lepidopterans in greenhouses, *Ostrinia nubilalis* Hubner in corn, and other Lepidopterans in orchards. The utilization of some microbial insecticides (such as those containing *Bacillus thuringiensis*) is also inundation.

Inoculative Releases

Inoculation involves releasing small numbers of natural enemies at prescribed intervals throughout the pest period, starting when the pest population is very low. The natural enemies are expected to reproduce themselves to provide more long-term control. However, the expected outcome of inoculative releases is to keep the pest at low numbers, never allowing it to approach an economic injury level; therefore, it is more of a preventive measure. However, the deference between inoculations and inundations is clear. A release with the expectation that the released organism will control the target after multiplication is inoculation. In glasshouses, the early release of parasitoids and predators, often with alternative food sources, is inoculation biological control. Examples of this are the releases of Encarsia formosa Gahan (Hymenoptera: Aphelinidae) and other natural enemies, now commonly practiced in glasshouses. The number of insects released is insufficient to control the pest insects, and success depends on the ability of the released organisms to multiply and reduce the target population. At the end of the season, the glasshouse is emptied, and no permanent establishment of the biological control organisms is achieved. When the next generation of plants is grown in the glasshouse, the predators and parasitoids should be released again. This type of release or application to control pest insects is strongly dependent on population regulation and density dependent processes.

9.9.2.3 Conservation Biological Control

Conservation biological control is defined as modification of the environment or existing practices to protect and enhance specific natural enemies of other organisms to reduce the effect of pests. Habitat manipulation often involves increasing the species diversity and structural complexity of agro-ecosystems. Habitat manipulation approaches provide natural enemies with resources such as nectar, pollen, physical refugia, alternative prey and alternative hosts and operate to reduce pest densities via an enhancement of natural enemies. However, although conservation biological control often increases natural enemy abundance, reduced pest abundance or increased yield has rarely been demonstrated. For example, flowering strips and other shelter habitats, as conservation biological control tactics, increase predation, parasitism, or yield in some cases but not others. In addition to natural enemies, conservation biological control tactics, such as habitat manipulation, attract and sustain a diverse suite of herbivores, detritivores, and plant provided foods. Moreover, research has been done on myriad arthropod pests, including species with high levels of insecticide resistance such as Chilo suppressalis (Lepidoptera: Crambidae) and Helicoverpa armigera -Lepidoptera: Noctuidae (Cory S. Straub *et al.*, 2007). As an example of conservation biological control, alternative habitats for natural enemies are provided, in the form of 'beetle-banks' in Britain or 'sown seed strips' in Switzerland in cereal crops. These practices are highly successful and are among the few documented uptakes of a biological control option in temperate open-field arable agriculture.

9.10 Biological Control Agents

Biological control agents or natural enemies are organisms such as insects or plants disease that are used to control pest species. Natural enemies of insect pests, also known as biological control agents, include predators, parasitoids, and pathogens. Biological control agents of plant diseases are most often referred to as antagonists. Biological control agents of weeds include herbivores and plant pathogens.

9.10.1 Predators

Predatory insects are beneficial because they feed directly on other insects like aphids. Common predatory insects include lacewings, ladybugs, and praying mantids. Ladybugs or lady beetles (*Coleomegilla maculata*) have been recognized by many cultures for their predatory behaviors for centuries. Adults of these insect predators are some of the most widely recognized insects in the United States. As early as the late 1800s, lady beetles were being used in biological control programs in the United States. Adult and larvae feed on large numbers of small, soft-bodied insects such as aphids but they will also eat other small, soft-bodied insect larvae, insect eggs, and mites (*www.gardeningsolutions.ifas.ufl.edu*). Example of predatory insect commonly found in garden is Green lacewing (*Chrysoperla carnea* Stephens) the adult, primarily feed on nectar and other fluids, but some species also consumes a few small insects. The larvae sometimes called aphid's lions; are voracious predators capable of feeding on small caterpillars and beetles, as well as aphids and other insects. They possess excellent searching qualities, exhibit high dispersal ability. Predatory stink bug, *Euthyrhynchus floridanus* (Linnaeus), is considered a beneficial insect because most of its prey consists of plant-damaging bugs, beetles, and caterpillars. Nymphs of the Florida predatory stink bug, *Euthyrhynchus floridanus* (Linnaeus) feeding on an earwig.

9.10.2 Parasitoids

The term parasite is frequently used for insects that parasitize other insects. Parasites are usually much smaller than their host and have a shorter life cycle than their host. Usually they do not kill their host. Examples include tapeworms and ticks. While a parasitoid is an organism that spends a significant portion of its life history attached to or within a single host organism in a relationship that is in essence parasitic. However, unlike a true parasite, it ultimately sterilizes or kills, and sometimes consumes, the host. When the parasitoid completes its life cycle, it becomes a free-living insect, no longer dependent on the host. Parasitoids may comprise up to 25 per cent of all insects (Parasitoids, Nick Mills, University of California, and Berkeley). Most parasitoids belong to the Hymenopterans or Dipterans orders. Parasitoid attacked different life stages of pest, *i.e.* egg parasitoid.

Similarly, egg parasitoids; are parasitoids that both attack and complete their development within a host eggs. They may be either solitary or gregarious, but in all cases prevent the host egg from hatching and use only a single host individual to complete their development. This distinguishes true egg parasitoids from other guilds of parasitic Hymenoptera such as egg-pre-pupal parasitoids that attack the host egg but delay development to kill the host just before pupation and egg predators that consume multiple eggs within an ovisac or egg pod. Example of egg parasitoid is Trichogramma Spp. It's used to control Corn borer (*Ostrinia* spp) in the farm, especially in developed countries like China and USA.

Dennise Medina and Arelys Rodriguez, 2010 categorized parasitoids with respect to its effect on host as; Idiobiont parasitoids are those that prevent further development of the host after initially immobilizing it, and, almost without exception, develop outside the host. Koinobiont parasitoids allow the host to continue its development while feeding upon it, and may parasitize any host life stage. In turn, Koinobiont can be subdivided further into endoparasitoids; which develop inside body of the host, and ecto- parasitoids; which develop outside the host body, though the parasitoids frequently are attached or embedded in the host's tissues.

9.10.3 Pathogens

Pathogens are microorganisms including certain bacteria, fungi, nematodes, protozoa, and viruses that can infect and kill the host. Populations of some aphids, caterpillars, mites, and other invertebrate are sometimes drastically reduced by naturally occurring pathogens, usually under conditions such as prolonged high humidity or dense pest populations. In addition to naturally occurring disease outbreaks, some beneficial pathogens are commercially available as biological or microbial pesticides. These include *Bacillus thuringiensis* or *B.t,* entomopathogenic nematodes, and granulosis viruses. Naturally occurring entomopathogens are important regulatory factors in insect populations. Many species are employed as biological control agents of insect pests in row and glasshouse crops, orchards, ornamentals, range, turf and lawn, stored products, and forestry and for abatement of pest and vector insects of veterinary and medical importance. The advantages of use of microbial control agents are numerous. These include safety for humans and other non-target organisms, reduction of pesticide residues in food, preservation of other natural enemies, and increased biodiversity in managed ecosystems.

Similarly, microbial control agents can be effective and serve as alternatives to broad-spectrum chemical insecticides. However, their increased utilization will require, increased pathogen virulence and speed of kill improved pathogen performance under challenging environmental conditions (cool weather, dry conditions, *etc.*); greater efficiency in their production; improvements in formulation that enable ease of application, increased environmental persistence, and longer shelf life; better understanding of how they will fit into integrated systems and their interaction with the environment and other integrated pest management (IPM) components; greater appreciation of their environmental advantages and acceptance by growers and the general public.

9.10.4 Entomopathogenic Nematodes

Nematodes are simple roundworms, colorless, unsegmented, and lacking appendages. They may be free-living, predaceous, or parasitic. Many of the parasitic species cause important diseases of plants, animals, and humans. Other species are beneficial in attacking insect pests, mostly sterilizing or otherwise debilitating their hosts. Entomopathogenic nematodes in the families' Steinernematidae and Heterorhabditidae have been used to suppress populations of insect's pest in a variety of agro ecosystems, and in several cases their positive effects on crop yield have been shown.

However, entomopathogenic nematodes (EPN) are mass produced for use as biopesticides using *in vivo* or in vitro methods. *In vivo* production (culture in live insect hosts) requires a low level of technology, has low startup costs, and resulting nematode quality is gene rally high, yet cost efficiency is low. *In vivo* production may be improved through innovations in mechanization and streamlining. A novel alternative approach to *in vivo* methodology is production and application of nematodes in infected host cadavers; the cadavers (with nematodes developing inside) are distributed directly to the target site and pest suppression is subsequently achieved by the infective juveniles that emerge. *In vitro* solid culture, *i.e.*, growing the nematodes on crumbled polyurethane foam, offers an intermediate level of technology and costs. In vitro liquid culture is the most cost efficient production method but requires the largest startup capital. Liquid culture may be improved through progress in media development, nematode recovery, and bioreactor design.

Furthermore, it has been reported that the beneficial entomopathogenic nematodes including *Steinernema carpocapsae* and *Heterorhabditis indica* were effective in infecting and killing soil-dwelling larval and possibly pupal stages of the striped flea beetle, *Phyllotreta striolata* and helping to reduce the emergence of future generation adult flea beetles from pupae. Similarly, the efficacy of EPNs against the cucurbit fly, *Dacus ciliatus* Loew (Diptera: Tephritidae) using high virulence native strain of *Heterorhabditis bacteriophora* Poinar (Rhabditida: Heterorhabditidae) and a commercial strain of *Steinernema carpocapsae* Weiser (Rhabditida: Steinernematidae). The efficacy of *S. carpocapsae* and *H. bacteriophora* was higher at 25 and 30°C than at 19°C. The results indicated that *S. carpocapsae* had the best potential as a biocontrol agent of *D. ciliatus,* based on its higher virulence and better ability to locate the fly larvae within infected fruits.

9.10.5 Entomopathogenic Fungi

Entomopathogenic fungi act as parasites of insects – these fungi can kill, or seriously disable insect pests. Fungal entomopathogens are important regulators of insect populations with considerable potential as mycopesticides. Only recently, however, have fungal entomopathogens been shown to occur as endophytes, both naturally and in response to various inoculation methods. The ecological function of endophytic fungal remains largely unknown, but some studies have implicated them in plant growth, herbivore resistance, and disease resistance. It is estimated that 750 to over 800 fungal species from more than 90 genera have been described as pathogenic against different insect species. But only a dozen of entomopathogenic fungus species are available for pest management at grower level.

Similarly, fungal biocontrol agents have unique mode of infection. In contrast to bacteria and viruses, they do not need to be ingested and can invade their host directly through the cuticle. That is why entomopathogenic fungi are capable of infecting non feeding mesh like eggs and pupae of insects. Moreover, fungal biological control agents have demonstrated efficacy against a wide range of insect pests including Spodoptera species. Application of entomopathogenic fungi against termites has the minimum negative impact on the environment. There have been a number of studies evaluating the efficacy of the hypocrealean Hyphomycete Beauveria bassiana (Bals.) Vuillemin. against subterranean termites. Lastly, the use of symbiotic fungal endophytes as biological control agents against fungal pathogens in cereals was first described in timothy (Phleum pratense L.) where plants containing the Epichlo typhina endophyte (E+) were less susceptible to disease by the fungus Cladosporium phlei.

9.10.6 *Bacillus thuringiensis* (B.t)

Bacillus thuringiensis (B.t) is a spore forming bacterium that produces crystals protein (cry proteins), which are toxic to many species of insects. *Bacillus thuringiensis* or B.t, is a naturally occurring soil bacterium that, when sprayed on plants, is toxic to certain pest insects. For years, farmers and home gardeners have used B.t as a microbial spray pesticide to control caterpillars, certain types of beetles, as well as mosquitoes and black flies. However, *B.t* occurs commonly in soils, and most insecticidal strains have been isolated from soil samples. Bacterial insecticides must be eaten by target insects to be effective; they are not contact poisons. Insecticidal products composed of a single Bacillus species or subspecies may be active against an entire order of insects, or they may be effective against only one or a few species. For example, products containing *Bacillus thuringiensis* var. kurstaki kill the caterpillar stage of a wide array of butterflies and moths. In contrast, *Bacillus popilliae* var. popilliae (milky disease) kills Japanese beetle larvae but is not effective against the closely related annual white grubs (masked chafers) that infest lawns in much of the Midwest. Indeed, commercial insecticides derived from this bacterium have a long history of successful use in the biocontrol of insect pests, in agriculture and forestry, and disease vectors. *B.t* Cry toxins constitute the active ingredient in the most widely used biological insecticides and insect-resistant transgenic crops.

However, Antagonistic bacteria such as *Bacillus licheniformis, Pseudomonas fluorescens* and *Bacillus* spp. or fungi such as Trichoderma harzianum or yeast such as *Saccharomyces cerevisiae* and *Rhodotorula minuta* have been found effective for the control of the anthracnose of mango under controlled research conditions. Another group of B.t isolates, including those from *Bacillus thuringiensis* var. *sandiego* and *Bacillus thuringiensis* var. *tenebrionis* are toxic to certain beetles. Within the order Coleoptera, species exhibit great differences in susceptibility to these isolates, presumably because of differences in the receptor sites in the gut wall of the insects where the bacterial toxins must attach.

Moreover, *B. thuringiensis* subsp. israelensis de Barjac (Bti) is used exclusively or in combination with other interventions for the control of larvae of dozens of species of medically important and pestiferous black flies and mosquitoes around the world. A prime example of the successful use of Bti occurred in the onchocerciasis control program in West Africa. High levels of resistance to the organophosphate insecticides that were originally employed for control of the Simulium vectors of onchocerciasis were threatening the future of the program.

9.10.7 Baculoviruses

Baculoviruses are a large group of double-stranded DNA viruses (almost 1000 species have been described); the majority have been isolated from a few insect orders: Lepidoptera, Diptera, Hymenoptera and Coleoptera. Viral genome ranges in size from 80 to 200 kb. Individual Baculoviruses usually have a narrow host range limited to a few closely related species. The most widely studied Baculovirus is the Autographa californica Nuclepolyhedron-virus (AcMNPV). Baculovirus insecticides have been used in a wide range of situations from forests and fields to food stores and greenhouses. Additionally, these viruses are excellent candidates for species-specific, narrow spectrum insecticidal applications. They have been shown to have no negative impacts on plants, mammals, birds, fish, or even on non-target insects. This is especially desirable when beneficial insects are being conserved to aid in an overall IPM program, or when an ecologically sensitive area is being treated.

However, Baculoviruses have several advantages over conventional insecticides that make them highly acceptable control agents. Probably the most important is their specificity. They have narrow host ranges (sometimes limited to one or two species), and do not infect beneficial insects, making them very suitable for use in integrated control programs. Most also possess the capacity to persist in the environment, which can be utilized in the development of more ecologically based long-term control programs. At present, commercial production of Baculoviruses has been carried out only *in vivo*, either by applying the virus against the host insect in the field and collecting diseased or dead larvae, or by producing the target insect in the laboratory on an artificial diet. The latter is the most commonly used method for producing Baculoviruses in many countries but both methods have been used successfully for commercial production of the Anticarsia gemmatalis baculovirus (AgMNPV) in Brazil.

Similarly, the activity of Baculoviruses against their natural hosts may be enhanced by introduction of insect-specific toxins or by interference with insect physiology. Most of the research was devoted to the studies of arthropod toxin genes isolated from the scorpion or spiders. The most potent insect-specific toxin gene used for construction of baculovirus recombinants was the gene coding for a toxin from scorpion *Androctonus australis*. The feeding damage caused by larvae infected with this modified baculovirus was reduced by about 60 per cent in comparison to a wild type baculovirus.

Chapter 10

Marketing and Exports of Organic Products

10.1 Introduction

Organic farming is one of the several approaches found to meet the objectives of sustainable agriculture. Organic agriculture offers trade opportunities for farmers in the developing and developed countries. This market of organic products is expected to grow globally in the coming years and high growth rates over the medium term (from 10-15 to 25-30 per cent) are expected. This organic market expansion makes it possible for farmers to reap the benefits of a trade with relatively high price premiums. However, this market is not very well known to most farmers, especially those living in the developing countries. Furthermore, information about it is not readily available to farmers in the developing countries. The absence of sufficient technical and market information and financial support also means that few farmers will risk changing their method of production. In developing countries it is therefore essential for major key players (*e.g.* NGOs, farmer organizations, traders, exporters *etc.*) that promote organic farming to have upto-date information on the available opportunities (market requirements) and trends of the organic market. One example is India, a country with a huge number of small farmers who still use traditional methods and do farming with few agricultural inputs. NGOs that promote organic farming and other organizations support farmers in these aspects. Niche markets have gradually been created, commonly based on trust and goodwill (formal certification did not begin until the 1960s and 1970s), and often using novel direct marketing strategies such as box schemes and community supported agriculture. After many years of consumers having to hunt around for their organic produce from several suppliers, perhaps directly from the farmer, the task is now a lot easier with specialist food shops and organic shelf space in supermarkets, in the industrialized world at least. Global links have been forged in all continents as organic agriculture has been seen to be an effective rural development option. Although the movement is still regarded with some skepticism the concept of organic farming has strong marketing appeal, growth forecasts are almost all positive and it has been suggested that the 'movement' is now an 'industry'. Organic agriculture is one of the fastest growing agribusiness sectors in the world, with double-digit annual grow thin land under organic cultivation, value of organic produce and number of organic farmers. An organic movement is now emerging in India on different levels (producer groups, trainers and advisors, certification bodies and processors and traders). So dissemination of information about the opportunities and challenges for Indian organic products on the domestic and international market is of fundamental concern in order to allow continued development of the organic agriculture movement in this country.

10.1.1 Market

The term "Market" is derived from the Latin word "Mercatus", meaning to trade. It may be defined in various ways; but in a simple way, "the market is a place where buyers and sellers converge". The buyer and seller meet and function; goods or services are offered for sale. And transfers of good's ownership takes place. Markets are primarily economic Institutions. It is through these institutions, the demand and supply forces get adjusted and monetization of the produces takes place. The concept of exchange leads to the concept of market. Marketing is a social and managerial process by which individual and groups obtain the material they need by paying the market value. This definition of marketing rests on the following core concepts: needs, wants and demands; products (goods, service and ideas); value, cost and satisfaction; exchange and transaction; relationships and networks; markets and marketers and prospects.

10.1.2 Classification of Market

There are different types of market. They can be classified under the following categories:

I. On Geographical Area Basis

- ☆ **Family Market:** During the period of village economy, these markets existed but are extinct now.
- ☆ **Local Market:** These markets existed during "town economy". They are gradually disappearing due to innovations and development in transport and communications. However, the local markets still exists in villages although their numbers are getting few. These markets are locally known by different names at different places. For example, in Assam, Uttar Pradesh and West Bengal, these markets are 'Haats' or 'Bazar' or 'Mandis' and in South India, they are called 'Mandis' or 'Shandis'.

- **National Market:** The rise and spread of industrialization has widened markets on the national level. Most of the products today have moved on to the national markets.
- **Urban and Rural Markets:** On the basis of location, markets are also classified into urban and rural markets. Urban markets are those located in cities or towns primarily catering to the need of city consumers.
- **World Market or International Market:** These markets came into existence with the growth of international transport and communication. The process of liberalization of world economy helped its consolidation. The world market for organic produce is now growing rapidly. It facilitates access to foreign goods/foreign currency for importers. The foreign market enables them to do transactions with the trade from another country. Often the market realization is better and remunerative in international markets.

II. On the Basis of Business

On the basis of business, there are mainly two types of markets:

- **Retail Markets:** The word 'Retail' comes from the French word 'retaillier' which refers to "cut into smaller sizes, clip or divide" in terms of tailoring. Retail market directly caters to the consumers who seek goods in smaller quantities. The retail markets include store retailing, non-store retailing and retail organizations.
  - **Retail Store:** There are eight categories of retail store *i.e.* Specialty Stores, Departmental Stores, Super Markets, Convenience Stores, Discount Stores, Off-price Retailers, Super Stores and Catalogue Showroom. The Super Markets play major role in marketing of organic products at the global level. These retailing stores are relatively large, low cost, low margin, high volume, self service operations designed to serve the consumer's total need. The advantages of these Super Markets are:
    - Large quantities can be sold.
    - Regular demand.
    - Possible high price level.
    - Access to large and growing market

    Various Departmental Stores in India also deal in organic produces. The Kendriya Bhandar located in different parts of the country is an example.
  - **Non-store Retailing:** Non-store Retailing falls into four major categories: direct selling, direct marketing, automatic vending and buying services. Now-a-days, non-store retailing has been growing much faster than store-retailing. In Aurangabad (Maharashtra), the

IIRD (Institute of Integrated Rural Development) has set up organic bazaars for direct selling of organic products to consumers. It has captured the attention from other marketing agencies from other parts of the State. Direct on-farm selling has following advantages:

- Direct contact to consumers.
- Building personal trust
- Higher profit by avoiding middlemen.
- Basics of Marketing of Organic Produce Independence.
- Permanent income

- **Retail Organization:** Retail organizations generally include corporate chain stores, voluntary marketing chains, retailer cooperatives, consumer's cooperatives and franchise organizations *etc*. These organizations have greater purchasing power, wider brand recognition and better trained employees. Reliance Fresh is one of the biggest retail organizations in India.

- **Wholesale Markets:** Wholesale market is the market where a wholesaler is the supplier and retailers are the buyers. Here goods are bought and sold in bulk quantities. Wholesalers (also called distributors) are involved in selling, promoting, buying, warehousing, transportation, financing, risk bearing, market information *etc*. In general, there are four types of wholesalers: merchant wholesalers, brokers and agents, Retailers branches offices and miscellaneous wholesalers.

10.2 Organic Market in India

India Organic Food Market Outlook. The India organic food market stood at a value of USD 849.5 million in 2020. The market is further expected to grow at a CAGR of about 20.5 per cent in the forecast period of 2021 and 2026 to reach a value of about USD 2601 million by 2026.

India is a leader in the export of **organic tea, basmati rice and cotton**. Another area where India can see a demand in the export market is through organic vegetables. The demand for organic fruit and vegetables is likely to continue right till the forecast year 2026.

In recent years, Indian consumers have started paying more attention to their health, the nutrient content and quality of the food they eat. Due to these concerns, they have gradually begun shifting towards organic food, even though organic foods are priced higher than conventional foods. This awareness among consumers is further giving a boost to the organic market as consumers are willing to pay for them if it is going to benefit their health and wellness. Also, in places where consumers are aware about the safety and quality of organic products, the market is growth is increasing, since consumers are willing to pay higher prices.

Organic farming in India is experiencing a real boom, when considering the number of farmers turning to organic practices. A large number of farmer groups, companies, NGOs, developmental agencies and government bodies promote organic farming in one way or the other. Although the organic market Basics of Marketing of Organic Produce is rapidly growing, the Indian organic sector is still scattered and its development lags far behind its actual potential.

The domestic market for organic products is yet to realize its potential. The products available in the domestic market in organic quality are rice, wheat, tea, coffee, pulses, fruit, spices and vegetables. Wholesalers/traders and Super Markets play major roles in the distribution of organic products. Most organic production originates from small farmers. Wholesalers/traders account for a 60 per cent share in the distribution of organic products. Large organized producers distribute their products through super markets as well as through self-owned stalls.

10.3 Channel of Distribution and Role of Middlemen

A channel of distribution (sometimes called a Marketing Channel) is a group of individuals and organizations that direct the flow of products from producers to customers. The channel of distribution includes the original producer, the final buyer and any middlemen-either wholesaler or retailer.

The term 'middlemen' refers to those individuals or institutions who facilitates, negotiate or selling in the capacity of an agent or broker. From the marketing point of view, the middlemen and intermediaries are the same. Most producers work with marketing intermediaries to bring their products to market. These marketing intermediaries generally make up a marketing channel.

The main players are:

- **Broker:** An intermediary whose job is to bring together buyers and sellers. The broker does not carry inventory, but often gets involved in finances or risk assessing.
- **Facilitator:** An intermediary who assists in the distribution process but neither takes title to goods nor negotiates purchase or sales.
- **Merchant:** An intermediary, who buys, takes title to and resells merchandise.
- **Producer:** An individual or a company that produces goods for self or market.
- **Retailer:** A business enterprise that sells produce/goods/services directly to the consumer.
- **Sales Agent:** An intermediary who searches for customers and negotiates on a producer's behalf but does not take title to the goods/produce.
- **Sales Force:** A group of people hired directly by a company to sell its products and service its accounts.

☆ **Wholesaler (distributor):** A business enterprise that sells produces/ goods or services to those who buy for resale or other business use.

10.4 The Organic Export Market

Global Market

Organic food market accounted for $162,036.7 million in 2019 and is predicted to generate a revenue of $416,049.7 million by 2027 with a CAGR of 12.4 per cent in the forecast period. North America market accounted for $59,305.4 million in 2019 and is predicted to rise with a CAGR of 13.1 per cent in the forecast period. Increasing demand for organic food among the population is predicted to drive the market in the estimated frame.

Organic food is being produced by various method of farming without using any preservatives or chemicals. The term organic farming is referred to farming by using various organic techniques in the farm. Organic farming involves the use of natural fertilizers like manure to improve plant growth. Moreover, animals raised organically are not given any hormones and are raised giving natural elements.

Recent years have seen very rapid growth in organic farming, particularly in Europe and the United States, but also in many other regions of the world including China, Latin America, and Africa. In the European Union, certified and policy-supported organic production accounted for just 100,000 ha on

Consumer demand for "healthier" food has been considered one of the major factors influencing increased consumer demand for organic food. This perception is probably reflected in the lower consumption of livestock products among the consumers most likely to buy organic food.

A survey of consumer choice in the United Kingdom showed that apart from health concerns, organic food was bought for environmental and animal welfare reasons. Approximately 70 per cent of consumers buying organic food mentioned these two factors.

Alongside the increase in the supply base, the market for organic produce has also grown. There has been a significant change in the development of the market for organic products, from supply-driven to demand led. In response, the role of multiples and multi-nationals in the processing, sale, and marketing of organic produce is increasing and the proportion or organic food purchased from specialist organic operations and local marketing schemes is decreasing as a result.

The trade in organic products is growing rapidly and becoming a reality throughout the world. Growth rates in the sector show organic products, that a few years ago supplied niche markets, have now entered mainstream marketing channels.

Many restaurant chefs across the country are using organic produce because they desire its superior quality and taste. Organic food is also gaining acceptance on a worldwide basis, with nations like Japan and Germany becoming important international organic food markets.

Though only a small percentage of farmers are expected to become organic producers, consumers demand for organically produced food and fibre products provides new market opportunities for farmers and businesses around the world. For many years, and with great success, the private sector alone has developed the concepts and markets for organic products. However, the surge in consumer interest has created new interest from the public sector, and developing countries are particularly in need of good information.

Table 10.1. State-Wise Export during 2020-21

Sl. No.	*State*	*Exported Qty (in MT)*	*Total Value (in Crore)*	*Total Value (in USD Million)*
1	Madhya Pradesh	500636.68	2683.58	394.64
2	Maharashtra	126272.80	913.74	134.37
3	Gujarat	65476.39	723.20	106.35
4	Kerala	8610.66	355.34	52.26
5	Haryana	38986.98	348.77	51.29
6	Karnataka	17436.02	321.36	47.26
7	Telangana	7532.38	280.96	41.32
8	Uttar Pradesh	12141.77	278.56	40.96
9	West Bengal	4642.18	269.97	39.70
10	New Delhi	31669.54	235.94	34.70
11	Daman and Diu	42404.62	231.62	34.06
12	Rajasthan	24713.51	217.95	32.05
13	Tamil Nadu	4223.70	108.74	15.99
14	Andhra Pradesh	1828.16	57.68	8.48
15	Goa	303.21	19.79	2.91
16	Uttarakhand	261.62	10.97	1.61
17	Jammu and Kashmir	607.59	10.77	1.58
18	Chhattisgarh	74.40	4.23	0.62
19	Punjab	326.03	2.73	0.40
20	Himachal Pradesh	7.88	1.62	0.24
21	Assam	22.89	0.75	0.11
22	Meghalaya	0.48	0.16	0.02
23	Jharkhand	0.20	0.07	0.01
	Total	**888179.69**	**7078.50**	**1040.96**

The export opportunities offered by the so-called developed world have clearly been a major stimulus in getting organic farming established in many countries. It is unlikely that organic agriculture would have developed so successfully in the northern hemisphere if there had been no demand pressure. Supplying a market in which demand outstrips supply is a producer's dream.

This has been the case with organic farming for a long time and figures show that this trend is set to continue. Yet, the sector has shown it is also vulnerable and that prices can collapse if there are sudden and exceptional increases in the supply of particular products. Therefore, marketing with a clearly targeted strategy is a must for organic products.

The organic market is a special one, partly because premium prices are involved. It requires a special effort. Attention should be given to meeting the requirements of a guarantee system that will ensure organic quality and allow consumers to develop their preferences for organic products with a feeling of trust.

The World Food summit Plan of Action recognized the importance of appropriate input technologies, farming techniques and other sustainable methods, such as organic farming, to assists farming operations to be profitable, with the goal of reducing environmental degradation, while creating financial resources within the farming operation.

10.5 Export Market

Indian organic producers and exporters are well aware of the demand for organic products in developed countries. Products available for the export market are rice, wheat, tea, spices, coffee, pulses, fruits and vegetables, cashew nuts, cotton, oil seeds and medicinal herbs.

The channels adopted for the export of organic products, except for tea, are mainly through export companies. Organic tea is produced by major well organized tea estates which are exporting tea directly.

In the case of other organic products, predominantly small farmers are involved in producing organic products. Hence, these products are exported through exporters.

Organic products are mainly exported to the following countries (in order of priority):

- Europe: Netherlands, United Kingdom, Germany, Belgium, Sweden, Switzerland,
- France, Italy, Spain;
- Americas: USA, Canada;
- Middle East: Saudi Arabia, UAE;
- Asia: Japan, Singapore;
- Australia;
- Africa: South Africa.

Table 10.2. Country-Wise Comparative Report of Organic Export-2020-21

Sl. No.	Country Name	2019-2020			2020-2021			Year on Year	Year on Year	Year on Year
		Exported Qty (in MT)	Total Value (in INR Lac)	Total Value (in USD Million)	Exported Qty (in MT)	Total Value (in INR Lac)	Total Value (in USD Million)	Growth Quantity	Growth Value	Growth Value (in USD Million)
1	U.S.A.	376070.03	240247.90	353.31	500935.95	379294.17	557.79	33.20	57.88	57.88
2	EUROPEAN UNION	175674.41	170670.46	250.99	267076.35	241957.12	355.82	52.03	41.77	41.77
3	CANADA	64225.71	31700.45	46.62	69142.41	38742.31	56.97	7.66	22.21	22.21
4	*GREAT BRITAIN	0.00	0.00	0.00	20844.14	13289.51	19.54	-	-	-
5	AUSTRALIA	2357.29	4754.29	6.99	2923.47	7401.82	10.89	24.02	55.69	55.68
6	SWITZERLAND	5192.01	6547.03	9.63	3924.64	5212.67	7.67	-24.41	-20.38	-20.38
7	ISRAEL	2531.74	1730.14	2.54	4480.12	3162.82	4.65	76.96	82.81	82.82
8	KOREA REPUBLIC	1292.16	947.77	1.39	4536.34	2931.31	4.31	251.07	209.28	209.25
9	U.A.E.	735.04	1432.34	2.11	1074.82	2308.47	3.40	46.23	61.17	61.21
10	ECUADOR	580.46	271.24	0.40	3708.05	1985.18	2.92	538.81	631.88	631.58
11	NEWZEALAND	1758.12	1224.54	1.80	1855.59	1851.59	2.72	5.54	51.21	51.19
12	JAPAN	1368.73	1834.08	2.70	611.32	1761.30	2.59	-55.34	-3.97	-3.97
13	VIETNAM	3270.25	1131.53	1.66	3276.59	1474.22	2.17	0.19	30.29	30.29
14	SRI LANKA	707.31	1037.34	1.53	300.13	928.34	1.37	-57.57	-10.51	-10.55
15	SAUDI ARABIA	162.14	372.24	0.55	325.67	584.78	0.86	100.86	57.10	57.22
16	SINGAPORE	390.16	582.82	0.86	348.05	513.29	0.76	-10.79	-11.93	-11.90
17	CHINA	436.23	615.20	0.91	238.29	494.76	0.73	-45.37	-19.58	-19.56
18	LEBANON	288.61	435.55	0.64	299.41	418.71	0.62	3.74	-3.87	-3.90
19	MALAYSIA	157.25	228.87	0.34	369.22	411.53	0.61	134.80	79.81	79.53

Sl. No.	Country Name	2019-2020			2020-2021			Year on Year	Year on Year	Year on Year
		Exported Qty (in MT)	Total Value (in INR Lac)	Total Value (in USD Million)	Exported Qty (in MT)	Total Value (in INR Lac)	Total Value (in USD Million)	Growth Quantity	Growth Value	Growth Value (in USD Million)
20	THAILAND	75.57	161.51	0.24	200.98	375.97	0.55	165.96	132.79	132.35
21	QATAR	252.97	255.42	0.38	241.22	326.34	0.48	-4.65	27.77	27.66
22	OMAN	176.77	273.52	0.40	209.31	269.13	0.40	18.41	-1.60	-1.49
23	KUWAIT	186.38	240.19	0.35	203.33	261.23	0.38	9.09	8.76	8.78
24	CHILE	60.70	47.79	0.07	87.95	209.18	0.31	44.89	337.69	340.00
25	BOLIVIA	154.00	96.89	0.14	218.98	196.22	0.29	42.19	102.52	103.52
26	BAHRAIN	120.56	131.55	0.19	150.22	179.87	0.27	24.60	36.73	37.31
27	MEXICO	240.93	375.34	0.55	39.00	178.30	0.26	-83.81	-52.50	-52.54
28	HONG KONG	40.77	83.25	0.12	72.40	171.56	0.25	77.58	106.07	106.56
29	MAURITIUS	27.80	63.51	0.09	94.19	157.30	0.23	238.81	147.67	148.39
30	EGYPT	4.59	26.13	0.04	25.33	140.85	0.21	451.85	438.93	444.74
31	SOUTH AFRICA	21.07	90.39	0.13	32.55	132.98	0.20	54.50	47.12	47.37
32	TURKEY	65.20	172.86	0.25	45.46	104.95	0.15	-30.28	-39.28	-39.37
33	FRENCH POLYNESIA	35.42	40.12	0.06	70.40	83.85	0.12	98.76	108.97	108.47
34	RUSSIA	105.20	68.48	0.10	60.61	53.99	0.08	-42.38	-21.16	-21.78
35	CAMEROON	0.04	4.17	0.01	29.69	52.44	0.08	84731.43	1158.11	1183.33
36	PHILIPPINES	43.05	22.75	0.03	19.66	41.27	0.06	-54.33	81.41	84.85
37	Sint Maarten	0.00	0.00	0.00	7.62	28.43	0.04	100.00	100.00	100.00
38	JORDAN	45.35	25.18	0.04	68.00	26.42	0.04	49.94	4.94	5.41
39	COLOMBIA	0.00	0.00	0.00	6.00	17.06	0.03	100.00	100.00	100.00
40	KENYA	0.00	0.00	0.00	0.13	16.47	0.02	100.00	100.00	100.00

Sl. No.	Country Name	2019-2020			2020-2021			Year on Year	Year on Year	Year on Year
		Exported Qty (in MT)	Total Value (in INR Lac)	Total Value (in USD Million)	Exported Qty (in MT)	Total Value (in INR Lac)	Total Value (in USD Million)	Growth Quantity	Growth Value	Growth Value (in USD Million)
41	NIGERIA	0.07	9.16	0.01	0.09	13.21	0.02	35.82	44.20	46.15
42	GHANA	2.97	4.19	0.01	3.39	12.47	0.02	14.47	197.78	200.00
43	PERU	0.30	4.68	0.01	2.86	12.20	0.02	853.33	160.74	157.14
44	BOSNIA AND HERZEGOVINA	5.00	9.14	0.01	5.00	11.75	0.02	0.00	28.60	30.77
45	GEORGIA	29.53	17.06	0.03	9.53	8.87	0.01	-67.73	-48.00	-48.00
46	MOLDOVA	0.35	14.96	0.02	0.34	8.67	0.01	-2.02	-42.08	-40.91
47	TRINIDAD AND TOBAGO	4.69	6.80	0.01	0.19	7.40	0.01	-95.89	8.84	10.00
48	UKRAINE	0.51	5.21	0.01	0.30	6.97	0.01	-40.86	33.69	25.00
49	ARGENTINA	0.00	0.00	0.00	1.85	4.11	0.01	100.00	100.00	100.00
50	NEPAL	32.01	38.67	0.06	0.35	4.08	0.01	-98.92	-89.45	-89.47
51	YOGOSLAVIA (SERBIA-MONTENEGRO)	0.00	0.00	0.00	1.20	4.23	0.01	100.00	100.00	100.00
52	SAINT LUCIA	0.08	2.08	0.00	0.14	3.49	0.01	76.92	67.80	66.67
53	BRAZIL	27.69	341.78	0.50	0.25	1.24	0.00	-99.10	-99.64	-99.60
54	INDONESIA	19.61	21.79	0.03	0.50	1.18	0.00	-97.45	-94.58	-93.75
55	MALDIVES	1.84	9.21	0.01	0.03	1.11	0.00	-98.37	-87.99	-85.71
56	ALGERIA	0.00	0.00	0.00	0.05	0.39	0.00	100.00	100.00	100.00
57	MOROCCO	3.09	44.29	0.07	0.03	0.41	0.00	-99.03	-99.09	-98.46
58	ARMENIA	5.00	4.98	0.01	0.00	0.00	0.00	-100.00	-100.00	-100.00

Sl. No.	Country Name	2019-2020			2020-2021			Year on Year	Year on Year	Year on Year
		Exported Qty (in MT)	Total Value (in INR Lac)	Total Value (in USD Million)	Exported Qty (in MT)	Total Value (in INR Lac)	Total Value (in USD Million)	Growth Quantity	Growth Value	Growth Value (in USD Million)
59	BARBODOS	0.00	0.00	0.00	0.00	0.09	0.00	100.00	100.00	100.00
60	BRUNEI	0.18	1.78	0.00	0.00	0.00	0.00	-100.00	-100.00	-100.00
61	CAYMAN ISLANDS	0.45	0.53	0.00	0.00	0.00	0.00	-100.00	-100.00	-100.00
62	GUADELOUPE	0.84	28.02	0.04	0.00	0.00	0.00	-100.00	-100.00	-100.00
63	ICELAND	0.20	3.54	0.01	0.00	0.00	0.00	-100.00	-100.00	-100.00
64	KAZAKISTAN	0.31	3.51	0.01	0.00	0.00	0.00	-100.00	-100.00	-100.00
65	PAPUA NEW GUINEA	0.00	0.12	0.00	0.00	0.00	0.00	-100.00	-100.00	100.00
66	TAIWAN	7.02	65.53	0.10	0.00	0.00	0.00	-100.00	-100.00	-100.00
67	TUNISIA	0.36	2.72	0.00	0.00	0.00	0.00	-100.00	-100.00	-100.00
68	UGANDA	1.50	1.03	0.00	0.00	0.00	0.00	-100.00	-100.00	-100.00
69	URUGUAY	0.82	7.20	0.01	0.00	0.00	0.00	-100.00	-100.00	-100.00
	Total	**638998.40**	**468590.82**	**689.10**	**888179.69**	**707849.52**	**1040.96**	**39.00**	**51.06**	**51.06**

*During 2019-20, the export to Great Britain had been covered in the export to European Union

Table 10.3. Category-Wise Comparative Report of Organic Export-2020-21

Sl. No.	Category Name	2019-2020			2020-2021			Year on Year	Year on Year	Year on Year
		Exported Qty (in MT)	Total Value (in INR Lac)	Total Value (in USD Million)	Exported Qty (in MT)	Total Value (in INR Lac)	Total Value (In USD Million)	Growth Quantity	Growth Value (in INR Lac)	Growth Value (in USD Million)
1	Processed Food	405383.99	214932.29	316.08	655986.21	400170.03	588.49	61.82	86.18	86.18
2	Oil Seeds	100815.43	62093.59	91.31	84072.66	63965.50	94.07	-16.61	3.01	3.01
3	Cereals and Millets	48677.64	38382.05	56.44	59907.79	51679.49	76.00	23.07	34.64	34.64
4	Spices and Condiments	8053.30	24369.01	35.84	10022.28	38061.41	55.97	24.45	56.19	56.19
5	Tea	6210.89	33800.64	49.71	6164.24	34512.04	50.75	-0.75	2.10	2.10
6	Medicinal Plant Products	2806.06	17999.76	26.47	4230.57	31935.76	46.96	50.77	77.42	77.42
7	Dry Fruits	3714.48	23372.05	34.37	3658.17	23848.95	35.07	-1.52	2.04	2.04
8	sugar	41940.89	18344.50	26.98	40541.51	19027.83	27.98	-3.34	3.73	3.73
9	PULSES	4829.61	5989.79	8.81	8781.97	12991.59	19.11	81.84	116.90	116.88
10	Coffee	4584.91	11212.62	16.49	4381.51	11097.37	16.32	-4.44	-1.03	-1.02
11	Essential oil	145.33	4744.08	6.98	219.31	7129.64	10.49	50.91	50.28	50.28
12	FODDER	4999.13	2244.63	3.30	6876.94	4187.90	6.16	37.56	86.57	86.58
13	Others	5429.05	6190.66	9.10	655.78	3328.76	4.90	-87.92	-46.23	-46.23
14	Fresh Fruits and Vegetables	511.29	1327.72	1.95	1404.32	2633.97	3.87	174.66	98.38	98.31
15	Flowers	346.55	2636.33	3.88	190.13	1575.52	2.32	-45.14	-40.24	-40.24

Sl. No.	Category Name	2019-2020			2020-2021			Year on Year	Year on Year	Year on Year
		Exported Qty (in MT)	Total Value (in INR Lac)	Total Value (in USD Million)	Exported Qty (in MT)	Total Value (in INR Lac)	Total Value (In USD Million)	Growth Quantity	Growth Value (in INR Lac)	Growth Value (in USD Million)
16	Tuber Products	495.95	848.58	1.25	1042.85	1504.50	2.21	110.27	77.30	77.32
17	Oils and Oleoresins	0.44	31.41	0.05	1.44	114.39	0.17	227.50	264.23	265.22
18	Miscellaneous	53.46	71.11	0.11	42.03	84.86	0.13	-21.39	19.34	19.05
	Total	**638998.40**	**468590.81**	**689.11**	**888179.69**	**707849.52**	**1040.96**	**39.00**	**51.06**	**51.06**

Table 10.4. Country-Wise Organic Export during 2020-21

Sl.No.	Country	Exported Qty (in MT)	Value (in Crore)	Value (in USD)
1	U.S.A.	500935.95	3792.94	557.79
2	European Union	267076.35	2419.57	355.82
3	CANADA	69142.41	387.42	56.97
4	GREAT BRITAIN	20844.14	132.90	19.54
5	KOREA REPUBLIC	4536.34	29.31	4.31
6	ISRAEL	4480.12	31.63	4.65
7	SWITZERLAND	3924.64	52.13	7.67
8	ECUADOR	3708.05	19.85	2.92
9	VIETNAM	3276.59	14.74	2.17
10	AUSTRALIA	2923.47	74.02	10.89
11	NEWZEALAND	1855.59	18.52	2.72
12	U.A.E.	1074.82	23.08	3.39
13	JAPAN	611.32	17.61	2.59
14	MALAYSIA	369.22	4.12	0.61
15	SINGAPORE	348.05	5.13	0.75
16	SAUDI ARABIA	325.67	5.85	0.86
17	SRI LANKA	300.13	9.28	1.37
18	LEBANON	299.41	4.19	0.62
19	QATAR	241.22	3.26	0.48
20	CHINA	238.29	4.95	0.73
21	BOLIVIA	218.98	1.96	0.29
22	OMAN	209.31	2.69	0.40
23	KUWAIT	203.33	2.61	0.38
24	THAILAND	200.98	3.76	0.55
25	BAHRAIN	150.22	1.80	0.26
26	MAURITIUS	94.19	1.57	0.23
27	CHILE	87.95	2.09	0.31
28	HONG KONG	72.40	1.72	0.25
29	FRENCH POLYNESIA	70.40	0.84	0.12
30	JORDAN	68.00	0.26	0.04
31	RUSSIA	60.61	0.54	0.08
32	TURKEY	45.46	1.05	0.15
33	MEXICO	39.00	1.78	0.26
34	SOUTH AFRICA	32.55	1.33	0.20
35	CAMEROON	29.69	0.52	0.08

Sl.No.	Country	Exported Qty (in MT)	Value (in Crore)	Value (in USD)
36	EGYPT	25.33	1.41	0.21
37	PHILIPPINES	19.66	0.41	0.06
38	GEORGIA	9.53	0.09	0.01
39	Sint Maarten	7.62	0.28	0.04
40	COLOMBIA	6.00	0.17	0.03
41	BOSNIA AND HERZEGOVINA	5.00	0.12	0.02
42	GHANA	3.39	0.12	0.02
43	PERU	2.86	0.12	0.02
44	ARGENTINA	1.85	0.04	0.01
45	YOGOSLAVIA (SERBIA-MONTENEGRO)	1.20	0.04	0.01
46	INDONESIA	0.50	0.01	0.00
47	NEPAL	0.35	0.04	0.01
48	MOLDOVA	0.34	0.09	0.01
49	UKRAINE	0.30	0.07	0.01
50	BRAZIL	0.25	0.01	0.00
51	TRINIDAD AND TOBAGO	0.19	0.07	0.01
52	SAINT LUCIA	0.14	0.03	0.01
53	KENYA	0.13	0.16	0.02
54	NIGERIA	0.09	0.13	0.02
55	ALGERIA	0.05	0.00	0.00
56	MALDIVES	0.03	0.01	0.00
57	MOROCCO	0.03	0.00	0.00
58	BARBODOS	0.00	0.00	0.00
	Total	**888179.69**	**7078.50**	**1040.96**

10.6 World of Organic Agriculture

As per the details released by BioFach 2010 at Nuremburg, the organic agriculture is developing rapidly, and statistical information is now available from 154 countries of the world. The main results of the latest global survey on certified organic farming are summarized below:

- ☆ 35 million ha of agricultural land is managed organically by almost 1.4 million producers.
- ☆ Regions with the largest areas of organically managed agricultural land are Oceania (12.1 million ha), Europe (8.2 million ha) and Latin America (8.1 million ha)

- Countries with the highest number of producers are India (340,000), Uganda (180,000) and Mexico (130,000)
- About one-third of the world's organically managed agricultural land (12 million ha) is located in developing countries.

10.7 Constraints and Opportunities for Organic Products

The demand for organic products has created new export opportunities for the developing world. While some consumers express a preference for locally-grown organic foods, the demand for a variety of foods year-round makes it impossible for any country to source organic food entirely within its own borders. As a result, many developing countries have begun to export organic products successfully (*e.g.*, tropical fruit to the European baby food industry, Zimbabwean herbs to South Africa, six African nations export cotton to the European Community). Typically, organic exports are sold at impressive premiums, often at prices 20 per cent higher than identical products produced on non-organic farms. The ultimate profitability of organic farm varies, however, and few studies have assessed the long-term potential for such market premiums. Nevertheless, under the right circumstances the market returns from organic agriculture can potentially contribute to local food security by increasing family incomes.

Farmers are denied access to developed country organic markets for two to three years after beginning organic management since such countries will not certify land and livestock as organic before that time, arguing that it is necessary for the purging of chemical residues. However, products produced on land under organic management for at least one year but less than the two-three year standard can be sold as transitional organic, although few markets have developed for such products.

In most cases farmers and post-harvest businesses seeking to sell their products in developed countries must hire an organic certification organization to annually inspect and confirm that these farms and businesses adhere to the organic standards established by various trading partners.

The cost for this service can be expensive, although it varies in relation to farm size, volume of production, and the efficiency of the certification organization (*e.g.* IFOAM certification costs a maximum of 5 per cent of sales value). Few developing countries have certification organizations within their borders, and even when sufficient resources are available to pay for certification farmers often lack the information to find credible inspectors.

While most developing country have focused on export markets in the developed world, domestic market opportunities for organic food or eco-food may also be exploited. In China, for example, there is a growing market for "green food" which, according to government grading standards, is produced without certain pesticides and fertilizers and with biological methods. Chinese farmers also produce organic food for export (*e.g.* tea to the Netherlands, soybeans to Japan).

Whether the intent is to sell organic products domestically or abroad, reliable market information is difficult to obtain. There is virtually no systematic production or market survey data being collected with which to assess the rate and pattern of organic market growth. In particular, no projections for the market in the developing world have been made, nor have markets systematically been identified for developing country exports. Estimates of the public's willingness to pay premiums, the impact of regional and tastes, and the incidence of market fraud have not been undertaken.

There is a growing consumer demand in the developed countries for healthy food. This change in the consumer preference reflects the increasing concern about the dangers associated with the consumption of food containing small quantities of food additives and pesticide residues. As the organically produced food is devoid of such harmful additives and residues, it is regarded as healthy food and, as such, enjoys special consumer preference. To the average consumer in the developed countries, organically produced food is more acceptable than the food produced under the conventional agricultural system because of the special qualities possessed by the former in regard to taste and nutrition.

In the developing countries organically produced food has not yet gained consumer acceptance to any appreciable extent. This is despite the presence of harmful chemical residues at higher concentrations in vegetables, fruits and other farm produce. In India, through one meal comprising of chapattis, rice, dal, vegetables, salad (mixture of onions, tomatoes, carrot and cucumber), curd and banana/orange or grapes as dessert, one consumers 0.5 mg of DDT and BHC plus some malathion and endosulphan. This is 40 times more than the amount what an average American or European consumers. These pesticide residues lead to all kinds of chronic diseases of the heart, brain, kidney and liver including cancer.

There is lack of awareness among the people of the developing countries about poisons in the food they are consuming daily. Even for those who are aware of the dangers of poison in food, they do not have easy access to organic products. Wherever such products are available, they are costlier than the conventionally produced items mainly because organic agriculture does not enjoy the benefit of subsidy. However, there are still parts of India and other developing countries where people have access only to organic (chemical-free) foods because agro-chemicals have not yet penetrated into interiors and hinterlands. Unluckily, this fortunate situation may not prevail for long as agriculture is being developed as an industry without regard to the carrying capacity of the natural resource base.

On the other hand, in the developed countries there is a rapid awareness with regard to organically grown and processed foods and related issues. In Europe and U.S.A. in addition to Reform Houses on also sees special counters in departmental stores exclusively for organic products. In the west, supermarkets are generally considered as a threat by environmental lists, but they may still have a role to play. By introducing a wholly different segment of the consumer population to organic foods, supermarkets can help increase sales of organic products, improve the consistency of supply by creating greater demand and increase overall customer awareness of organic issues. These supermarkets can plough the vast masses of consumers, turning non-users into light user, and light users into regular customers of organic specialty stores. A problem for specially stores and supermarkets is consumer resistance to the higher price of organic products. In spite of this, consumers who are aware are willing to pay higher price for such products not only because they taste better, but also because of the realization that, in the process, they are supporting organic farming and thus protecting the environment.

Organic agriculture is not only leading the way to viable alternatives to modern development models, but it is also part of the solution to global environment problems. Now, more than ever, opportunities exist for the organic market to expand from less than 1 per cent of overall food sales to between 10 and 20 per cent by the turn of the century. There are, however, critical and sometimes very sensitive issues to be addressed in order to sustain a growing market for organically produced food.

Unfair Trends in the Market

Over the last ten years organic products have been marketed in increasingly larger quantities outside the area in which they are grown. Whilst seasonal production and regional markets remain an important objective in organic farming, there is nevertheless plenty of export opportunities for such products from the South as coffee, tea, cacao, bananas, spices, herbs and other subtropical and tropical products. 'Ecological dumping' is becoming more common and is a practice strongly resisted by the organic movement. There is a very real difference between production in keeping with the holistic principals of the organic movement and the purely commercial production of 'natural', 'biological', or 'organic' products such as organic cash-crops in monoculture. It is not enough to exclude what is known as 'ecological dumping' from the organic sector. Given the holistic nature of our movement, we also care about social aspects. In joining forces with the Fair Trade movement we also hope to be able to seriously challenge the 'social dumping' that involves the exploitation of cheap labour and child labour.

Bio-colonialism is one of the major challenges facing the organic movement. Given the much greater economic wealth and 'buying power' of the North, many products of organic quality find their way into these premium markets and so become unavailable to local and regional populations. Much more effort and creativity is needed if organic food is to become accessible to more people throughout the world. The SEKEM initiative is a good example of how this can be achieved.

Community supported agriculture with direct links between producer and consumer, in Japan are Asia known as TEKEI, offer a way of getting out of the bio-colonialism trap. Today, there are organic shops in India and Malaysia, farmer markets in Brazil, and organic produce in supermarkets in Argentina. Box schemes and home delivery of organic products are also being set up in several southern countries. Very often organic farming systems begin with the production of 'cash crops' and gradually, with the knowledge gained in producing for export, the whole farming unit is converted to organic systems. Thus, planting shadow trees in coffee plantations and using compost is often followed by introducing organic methods into the cultivation of staple foods such as corn and beans. There has always been agricultural trade. Trade remains a necessity whether it is barter between neighbours or long-distance trade facilitated by money. Trade makes it possible for house-holds to supplement their own production and satisfy their need for food and money whilst exploiting comparative advantages.

In this century, there has been an enormous expansion in trade supported by cheap fossil energy, agr-chemicals, refrigeration and increasingly sophisticated forms of transport. At the same time, the buying power of urban populations in both the North and South have combined with trade liberalization and globalization to give a powerful stimulus to markets every-where. This trade expansion has brought benefits to many agricultural producers and consumers. However, it has also been responsible for an uncontrolled concentration of power and wealth in the hands of a few countries, companies and persons, most of whom can be found in the industrial West. Control over capital and knowledge-intensive technologies, unfair trade practices including the dumping of cheap subsidized food and the manipulation of small farmers whose market choices are already limited have seriously effected small farmers everywhere. At the same time the capacity of industrialized countries to protect their own agricultural markets and the worsening terms of trade between agricultural and industrial products have also contributed to the decline and poverty of many agricultural communities. As small, market-oriented farmer lose out to larger farmers and the non-farming professions, and trans-national companies gain control over international commercial flows and agricultural production, many leave their land to search for jobs elsewhere. This not only leads to an increasing gap between rich and poor but also to the loss of food security, ecological degradation, pollution and the decline of indigenous knowledge and cultural diversity. Market-oriented agriculture and agricultural trade can lead to a serious depletion of soil fertility, soil erosion and a loss of biodiversity. Agriculture under such circumstances is not ecologically sustainable. However, agricultural trade and market-oriented agriculture do not have to be a trap for the small farmer. Fair and green trade offers important economic and ecological opportunities for sustainable development.

Fair Trade

Alternative approaches are being developed in reaction to the free trade model. Since the 1970s, the Fair Trade movement has been developing fast. Important organizations buy direct from Southern producer partners, pay a fair price and establish long-term working relationship. Initially, health food and Fair Trade shops were important alternative trade outlets.

Today, however, Fair Trade products can be found in most supermarkets. Trade organizations that respect the principals of Fair Trade and accept external monitoring by labelling organizations, are awarded a seal of approval. The fundamental principal of Fair Trade is equal partnership between Southern producers, Northern importers, labeling organizations,

Fair trade is always associated with production in the developing world. However, we need fair and real cost covering prices for farmers all over the world. And here the organic market has something to offer all farmers. Organic production often requires more effort and may result in lower ('optimum rather than maximum') yields. Also, as millions of consumers all over the worlds have come to realize – 'nature has its price'.

Therefore, growing numbers of consumers are prepared to pay premium prices because they know they are getting premium quality products in return. In calculating the price of organic products more factors are involved than in the pricing of conventional products. Demand, for example, plays a much bigger role.

The complex of aspects involved makes it difficult for us to discuss price finding and setting mechanisms in the organic sector here. However, we can say that premium prices ranging from 20 per cent to 30 per cent seem to be normal and when combined with often higher fair trade premiums, they make it possible for small farmers to continue to work their land, feed their families and enjoy real development.

Fair Trade and Trade Development

'Fair trade' covers a wide range of ideas. Fair trade is seen as a conscious decision to help small producers in the Third World become viable exporters of their own produce and to support their learning process. Fair Trade does not mean, however, that other forms of trade are unfair. Fair trade, of course, is not agricultural commodities and agricultural producers. All disadvantaged producers of any marketable product fall within the definition of the fair trade movement. In practice the fair trade labelling movement and organizations such as Equal Exchange in the USA and ATJ in Japan are mainly concerned with coffee, cocoa, tea, bananas, honey and sugar. The older alternative trading organizations work on a smaller scale and deal with thousands of food and non-food products. In this article we focus on small, co-operatively organized, agricultural commodity producers.

Small Farmers Disadvantage

Major international agriculture commodities are frequently produced by small farmers who have no secure title to land. Their land holdings are small and they are unable to assert their civil and economic rights even when these are protected by law. Their access to market often depends on long and exploitative chains of intermediaries. Not only do they lack bargaining power, they have no access to information on world market prices. It is difficult for small farmers to acquire the technology and technical skills and the cost of financial services are frequently beyond their reach. As many farm in fragile, near subsistence economies, their choices of survival strategies are not necessarily rational from the economic and ecological point of view. Due to imperfections in the market, lack of transparency and politically inspired distortion, small farmers tend to receive lower prices for their produce.

Dilemma

There are many reasons why the problems of small farmers should be addressed. First, maintaining and boosting the economic viability of smallholdings puts a brake on migration to overcrowded cities incapable of offering employment alternatives. Second, creating an effective economic and institutional environmental makes it possible for people to exploit the market potential of their area. Finally, supporting farmers' initiatives by encouraging co-operatives and associations enables them, collectively, to gain fair access to the (world) market: trade is better than aid. In recent years, fair trade labelling has increased the effectiveness of such interventions.

Fair Trade Labeling

Fair trade labelling aims to develop market niches where newcomers to the world market can learn the ins and outs of this type of trade in a relatively protected environment. In doing so, it is tries to provide small farmers' organizations with fair access to the (world) market, under conditions appropriate to their needs. The movement also supports farmers' organizations in their attempt to secure working capital and long-term loans for investments, market information, communications, management support and technical assistance.

Promoters of Fair Trade

Over the years, alternative trading organizations have promoted fair trade through specialize chains of (Third) World shops. The movement began about 30 years ago in the Netherlands and the Fair Trade Organization is still one of the most dynamic organizations in the field. Today, trading organization are found in almost all industrialized countries and most of them share supplies' databases. In the eighties, the movement stagnated World Shops could not cope with the demand for markets coming from Third World suppliers. Mainstream distribution networks had to be accessed and pressure built up to professionalize the network of World Shops. This lead to the development of fair trade certification and tested in the model was first developed and tested in the Netherlands under the name Max Havelaar. Later is spread within and outside the European Union until today certification and labelling is carried out in some 15 countries. Some use the name 'Max Havelaar', others have chosen names and labels that reflect their own national and cultural identity. However, all are members of Fair Trade Labelling International. Coming up to standard

Action for Fair Trade

Fair trade organizations have been working with Southern producer organizations for decades to try and ensure that they receive a decent income for their work. Only recently, however, have fair trade products begun to reach mainstream Northern markets. Coffee is probably the most successful fair trade product. It is also the most important traded commodity after oil and the main export for many Southern countries. Most comes from family-owned farms in developing countries. Fair trade organizations work with the producer organization, promote the sale of fair trade coffee, monitor buyers and roasters and guarantee that labeled products meet fair trade criteria. The ethical trading organizations bear the costs of inspecting producers. The fair trade market was boosted by the development of recognizable fair trade marks awarded for products that met clear criteria. However, it was the development of a clearly recognizable brand. Cafe-direct, that brought fair trade coffee into mainstream retail outlets and within easy reach of consumers.

Cafe-direct was developed by a partnership of four fair trade organizations that bought coffee from 14 producer organization in Africa and Latin America. Sales have increased dramatically over the last five years and Cafe-direct controls 3 per cent of the roast and ground market and 2 per cent of the instant market in the United Kingdom despite being 10 per cent more expensive than most conventional brands.

As well as higher income and greater security, fair trade can also lead to environmental benefits. Fair trade can also lead to environmental benefits. Fair trade organizations often work with producers to improve the ecological sustainability of production.

Progress in Fair Trade Marketing

Fair trade certification has helped bring fair trade products into mainstream European distribution networks. Sales of coffee and bananas have increased tenfold. Today, small farmers are able to sell 14,000 tons of coffee with a trade value of some US$ 40 million in this way. In the Netherlands the market share of fair trade coffee has risen from 0.2 per cent to 2 per cent and in Switzerland to 5 per cent, Mechanisms and alliances have been created to make substantial amounts of short and medium- term ethical bank loans available. While the impact to fair trade labeling appears to be unquestioned, its global impact should not be overestimated. In a global perspective, fair trade is no more than a tiny niche market. Although its impact is growing, results have been reached under very special market conditions and at a time when world market prices were persistently below subsistence level. Under these circumstances, fair trade offered a valuable alternative to those who had their name on the Fair Trade Labelling Procedures 'Register and who had secured an outlet for their product. However, not all organizations used (were able to use) this opportunity to effectively develop the capacity to survive under 'normal' market conditions. In 1994, coffee prices on the world market started to return to 'normal' levels, *i.e.* above the floor price established by the fair trade labelling organizations. For many farmers' organizations, the price premium of US$ 0.05 per pound above the market price was insufficient, while the inflexibilities inherent in the fair trade model eroded a substantial part of the price premium.

A growing number of farmers' organizations who had been operating within the fair trade market ran into serious trouble despite higher-than-market prices, while others managed to survive and went onto develop into strong and went on to develop into strong and respected co-operative organizations, addressing the economic needs and rights to many thousands of small farmers and their families.

Protectionism

Unless alternative approaches to trade become mainstream, the inequity and environmental damage inherent in conventional trading practices will remain. There is need for a more appropriate economic framework to redirect the world economy towards sustainability.

This is not a feasible option in the short-term. A more immediate approach would be a new type of protectionism in which governments formulate trade regulations based on socially fair and ecologically sound standards capable of creating enabling conditions for green and Fair Trade.

Priority to Local Economics

Even if such legislation were to be adopted, it would not be possible to eliminate completely the negative social and environmental effects of international trade. There will always be an important group of farmers who are unable to compete on the (inter) national market and whose farms fall in the margins of the globalizing economy. International trade is wholly dependent on fossil energy, itself an important source of pollution and, in this sense, International Trade is inherently unsustainable. To avoid the environmentally unsound transport of agricultural products over long-distances and a risky dependence on distant consumers, sustainable development and protection of the local economy. Particular attention should be paid to marginalized peoples and guaranteeing ecological and cultural appropriateness. Strengthening the local economy fosters a community's sense of social and environmental accountability, self-reliance, food security, cultural and biological diversity and the commitment to conserving its natural resources base. Indigenous people in the Andes protect the cultural, ecological and economic integrity of their communities by sustaining barter – which as a system is probably more widely spread than officially recognized – as an important element of their economic and cultural system.

Strengthening Local Economics

In many communities economic activities function at an unnecessarily low level as money and local natural resources leak away into the wider market economy. Local exchanges of goods and services can be greatly enhanced by the development of local money systems. Local money system has successfully stimulated the local economy in Argentina. Many local groups and communities in the North and in the South are currently experimenting with these Local Exchange and Trade Systems (LETS).

Critical Factors

It would seem that in some cases fair trade procedures have not been adapted to the operational needs of farmers' organizations, especially under present unstable market conditions. In addition, the fair trade labelling organizations have not paid sufficient attention to the development and organization of the necessary management support services. Management is also a critical factor at farmer level. The minimum scale of operations needed to run an export business is far beyond the scope of small and often isolated farmers. Lacking the management skills needed to operate the business, they have to hire in managers over whom they may have little control.

If there is no effective control and accountability is not enforced, the temptation to address private interests above co-operative interests can be irresistible and leads to serious problems. In this way a vicious circle develops and the co-operative is excluded from credit and loans because it has become un-creditworthy and its commercial liability has been further compromised by inadequate quality control. Many co-operative organizations, however, have successfully exploited the opportunities of the fair trade market and built up and diversified their businesses to the benefit of their members. Critical success factors seem to be the availability of highly component and reliable management, sustained management support, and market diversification into conventional, fair trade and organic markets.

10.8 Challenges

There are several major challenges facing the fair trade and organic market. A strategy must be developed to facilitate the merger of ecological and fair criteria incorporating social fairness and ecological responsibility as preconditions for sustainable production. Criteria must be adapted to the operational needs and limitations to Third Worlds realities and international markets.

At the same time the management and information gap must be closed and a solution found to the recurrent problem of credit. It is clear that the certifying organizations cannot solve these huge problems on their own. Strong global alliances with trade and industry, development organizations, ethical (and conventional) financial institutions, universities and consultants are needed in order to build a system that works and hence challenges 'business as usual'.

Trade Opportunities

In today's globalize world, export success is one of the major routes to economic progress for developing countries. But the conditions for success are changing as producers face rising environmental expectations in key export markets, resulting from tightening regulations, new corporate practices and changes in consumer values and lifestyles.

There new expectations reflect the growing recognition that current patterns of consumption, particular in the richer, industrialized world, are not environmentally sustainable. Profound changes in the ways in which goods and services are produced, traded and consumed will be required, both to reduce the burden on the global environment and to ensure that a growing population has resources to meet its needs. For these who can adapt to these requirements and start moving to anticipate trends, there are new opportunities to be found in sustainable trade that can generate financial, environmental and social benefits. Already growing numbers of farmers in the Sought are receiving higher prices and more long-term security by selling their products into environmentally friendly or fair trade markets in the North. Consumer demand for organic products is gradually increasing in responding to concerns about the environmental and health implications of industrial agriculture.

New Opportunities in a Growing Market

Latin American producers have been quick to tap into these markets. The organic sector in Mexico, for example, is now estimated to be worth US$ 500 million and Argentina is making serious efforts to development its organic sector with sales rising from US$ 1.5 million to US$ 20 million in the last six years. This rapid development has been supported by efforts to overcome the bureaucracy surrounding EU recognition imported organic goods. Argentina became the first developing country to obtain a place in the EU provisional list, thereby gaining market advantages. New products have no established market; emerging markets suffer from either insufficient supply or over-supply of products. This contradiction also rules emerging markets for eco-friendly produce, necessitating a balancing act between supply and demand. Investments have to be made; initial losses, though inevitable, need to be pegged down, through judicious promotional campaigns and production planning, to avoid a demand-supply gap.

No compromise is possible regarding timely production targets, both qualitative and quantitative, to satisfy and maintain the trust of an emerging clientele. Production has to be so organized that producers are able to conventional inputs. Here the issues are the short shelf-life (of eco-friendly inputs such as bio-fertilizers and bio-control agents which are living organisms) and the crop-specific or pest-specific action of many of these. The fact that these inputs are in great demand and in large quantities, but only during specific times -grow in season – complicates production planning and marketing. Third, being eco-friendly also implies minimizing environmental damage in the course of transport and storage of the produce. Since energy-intensive operations have to be curtailed, transport and cold storage of eco-friendly foods for the right market become controversial. To overcome such dependence on non-renewable energy, production systems tuned to local markets and seasonal products may be well worth exploring. However, wherever local production potential is limited, the required product range, be it for a healthy diet or a healthy, has to be met through export from high-potential areas.

Alternative Markets

Experiments to develop parallel markets have shown that it is not easy to do so. The problems encountered range from regularity of supplies and storage problems (space, shelf-life and eco-friendly storage methods) to retaining the goodwill of consumers. The tiniest upset can result in clients dropping out. There are several realities about alternative marketing that are gradually coming to light.

Role of the Trader

It is almost impossible for a small organic farmer to sell all his/her produce in the village. This calls for a whole chain of arrangements, including the presence of a trader. Fair trade considerations mean that a balance of power has to be ensured between producer and trader. To achieve this, producers' collective must be organized. More often than not, in our villages, the trader is a key socio-economic actor – as money lender, input supplier, transport provider and linking agent with the larger market. By providing flexible and timely loans, assured transport and access to markets, he becomes an ally of small-scale producers, despite their indebtedness to him. He influences farmers' crop choices by being their financier and by sharing his intimate knowledge of markets; in turn he also creates a system for loan recovery.

Quality Guarantee

Any opportunity to earn quick money can increase the chances of producers and marketers cutting corners. It is not uncommon to come across organic produce with pesticide residues eco-friendly inputs that are either spurious or have crossed the prescribed expiry date being marketed. Such problems arise because of some unethical elements operating among producers and traders. Hence, quality guarantee through inspection and certification by independent bodies becomes inevitable. Besides quality, fair trade also implies that profit margins in the production-trade chain need scrutiny and certification. Who should pay for inspection and certification? Reasonably speaking, the producer, trader and consumer should share the expenses. For effectiveness, a local-level certifying body could be constituted and governed by representatives from organizations of producers, traders and consumers. It could assume responsibility for defining standards, checking quickly, and even promoting the use of eco-friendly produce. To bolster this, the retail outlet should use an identifiable tag to indicate the marketing of certified produce.

10.9 Significance of Organic Agriculture for Indian Farmers

In the Indian context, organic farming can be significant in two distinct ways:

1. ***To increase the efficiency and sustainability of production:*** Organic farming can help to reduce production costs (especially where labor is cheap compared to input costs) and to increase or stabilize yields on marginal soils. This is especially relevant for small holders in marginal areas where Green Revolution agriculture has lead to a depletion of soil fertility and to high debts because of increase in input costs.

2. ***To Increase Product Value:*** In areas where farmers have access to established organic markets within the country or abroad, products can achieve a higher price compared to the conventional market. Especially in the trend of decreasing prices for agricultural products, this can be an important way to stabilize or even increase incomes.

High Cost of Organic Food is Predicted to be the Biggest Restraint for the Organic Food Market

Production cost of the organic food is typically higher because of labor inputs and greater diversity of enterprises. This increases the eventual retail price, which is considered to be the biggest restraint for the global market in the forecast period. Moreover, the shelf life of organic food is very low, which is predicted to hamper the market growth in the estimated period.

Organic Food Market, by Food Type

Fruits and vegetables segment is predicted to be the most lucrative segment

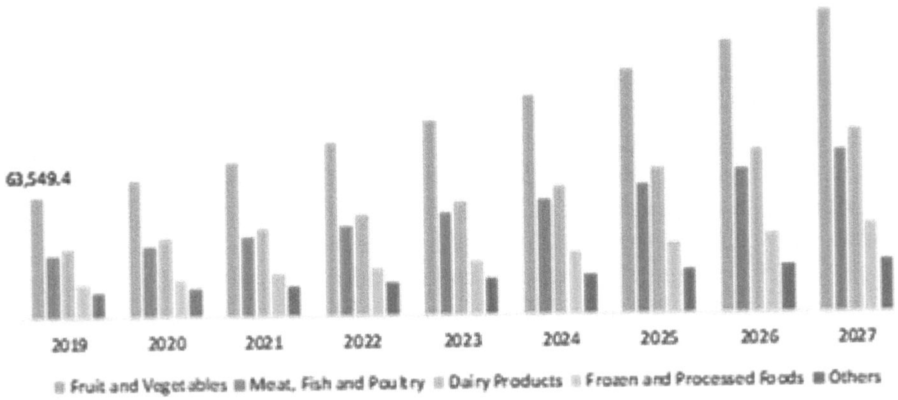

***Source*: Research Dive Analysis.**

Fruits and vegetable sub-segment is predicted to have the highest market share in the estimated period. Fruits and vegetable accounted for $63,549.4 million in 2019 and is predicted to rise with a CAGR of 12.1 per cent in the forecast period. Organic fruits and vegetable production prohibits the use of chemical pesticides, chemical fertilizers, or chemical preservatives. Increasing cultivation of various organic fruits and vegetables with the increase in the demand for various organic products across the globe is predicted to boost the segment in the estimated period.

Indian domestic market and export Organic Food Consumption in India is on the Rise. Some people believe that organic food is only a "concept" popular in the developed countries. They think that when it comes to organic food, India only exports organic food and very little is consumed. However, this is not true. Though 50 per cent of the organic food production in India is targeted towards exports, there are many who look towards organic food for domestic consumption. The most important reason for buying organic food is the concern for the health of children, with over 66 per cent parents preferring organic food to non-organic food. Though organic food is priced over 25 per cent more than conventional food in India, many parents are willing to pay this higher premium due to the perceived health benefits of organic food. The increase in organic food consumption in India is evident from the fact that many organic food stores are spurring up in India. Today every supermarket has an organic food store and every large city in India has numerous organic food stores and restaurants. This is a huge change considering that the first organic food store in Mumbai was started in 1997. Organic food exports from India are increasing with more farmers shifting to organic farming. With the domestic consumption being low, the prime market for Indian organic food industry lies in the US and Europe. India has now become a leading supplier of organic herbs, organic spices, organic basmati rice, *etc*. The increasing demand for organic food products in the developed countries and the extensive support by the Indian government coupled with its focus on agri-exports are the drivers for the Indian organic food industry. Organic food products in India are priced about 20-30 per cent higher than non-organic food products. This is a very high premium for most of the Indian population where the per capita income is merely USD 800. The domestic market is not sufficient to consume the entire organic food produced in the country. As a result, exports of organic food are the prime aim of organic farmers as well as the government.

Products for which Indian production has a comparative advantage:

India being a country with different agro-climatic zones, each state produces its own specialty products. Products for which production in India has a comparative advantage are given in the Tables 10.5 and 10.6 (Source: Org-Marg.)

Table 10.5

Product	*Season*	*States*	*Major Locations*
Cotton	Kharif	Maharashtra, Gujarat, M.P.	Akola, Amravati, Amreli, Kheda, Indore
Tea	Throughout the year	Assam, West Bengal, Uttarakhand	Darjeeling, Guwahati, Dehradun
Spices	Throughout the year	Kerala, Tamil Nadu, Karnataka	Cochin, Coimbatore, Idduki, Coorg
Coffee	Throughout the year	Kerela, Tamil Nadu, Karnataka	Coimbatore, Coorg, Wayanadu, Peeremade
Rice	Kharif and Rabi	Punjab, Haryana, Assam,Maharashtra, Tamil Nadu	Amritsar, Jalandhar, Darrang, Ratnagiri,Kanchipuram, Thiruvallur
Wheat	Kharif and Rabi	Punjab, Haryana, U.P.	Ambala, Patiala, Bhatinda, Faridkot
Vegetables and fruits	Throughout the year	All India	Various locations

Table 10.6. Obstacles and Solutions for the Low Import of Indian Organic Products

Aspect	*Obstacle*	*Solution*
Price	"Price expectations are too high in relation to quality"	The export traders from India should have realistic prices
Quality	"Low consistency of quality; contamination	The quality must be consistent. For this reason, higher quality standards must be enforced to develop and maintain a good reputation. Post-harvest practices should be improved.
Availability	"Reliability of exporters"	More promotion from Indian traders and governmental institutions. Better understanding of the demands of buyers (small quantities). Suppliers must show financial stability.
Logistics	"Slow shipment, restrictions for importing Indian organic products"	The logistics must be better coordinated from the place of dispatch for shipment. Duty ports must reduce the time taken to complete the customs process. Enhance the infrastructure available to guarantee quality upon arrival (cool rooms).
Certification	"Lack of national certification and accreditation	Local certification bodies accredited by national and international organizations would enhance the acceptance of organic products from India.
Export authorities	"Time consuming and complicated paper work."	Develop a fast track for export.
Information	"Lack of information on availability and certification"	More promotion activities on the part of traders, farmers and governmental institutions (*e.g.* international fairs).Establish an internet portal to enable easy access to information on Indian organic products.
Administration	"Inconvenient modus of payment (letters of credit); The governmental bank's system is too slow	Indian traders must show more flexibility in modus of payment. The bank's systems must reach higher standards.
Customer service	"Poor customer service from the Indian traders after sales."	Increase service quality; in particular, client follow-up systems must be implemented. Traders must accomplish what they promise.

Source: FiBL, 2002

Chapter 11

Biopesticides

11.1 Introduction: An Overview

Biopesticides are the biological agents or plant-based products used to control the population of injurious organisms to the ecosystem.

They are naturally occurring substances from living organisms (natural enemies) or their products (microbial products, phytochemicals) and their by-products (semiochemicals) that can control pests by non-toxic mechanisms. They are considered as minimal risk products safe to human and his environment.

It includes the use of botanicals, microbial pathogens such as fungi, bacteria, viruses and natural enemies of pests such as parasitoids and predators, nematodes and semiochemicals.

Indiscriminate use of synthetic pesticides resulted in the development of resistance, resurgence and outbreak of secondary pests. Stringent regulation in the use of synthetic pesticides and the demand for organics promote the use of biopesticides.

Biopesticides are the biological agents used to control the pest population. It includes the use of botanicals, microbial pathogens such as fungi, bacteria, viruses and natural enemies of pests such as parasitoids and predators, nematodes and semiochemicals. Biopesticides play an important role in sustainability of agricultural bioeconomy. The ecosystem benefits rendered by the agriculturally important biological resources warrant inclusion of biopesticides in Integrated Pest Management Programmes.

Management of pests without using harmful chemicals to human health and the environment is possible only by using biopesticides. Biopesticides play an important role in sustainable agriculture. It causes mortality by substantially reducing arthropod populations before it reaches the economic threshold level.

There is an urgent requirement to include biopesticides in Integrated Pest Management Programmes (IPM) to make it more sustainable since it lacks phytotoxicity, leaves no residues and is eco-friendly.

11.2 Biopesticides: Global and Indian Perspective

The contribution of biopesticides to insecticide sector accounts for 15 per cent of the total insecticide market. The estimated share of biopesticides in the world market is around $3 billion accounting for 5 per cent of the total crop protection (Marrone, 2014). The growth rate of a biopesticides is increased to 15 per cent. In a successful natural control program, only 15 per cent of biocontrol agents for control of insect pests have been identified. Commercially more than 125 species of natural enemies are available as biopesticides. Almost 90 per cent of the microbial biopesticides currently available in the market are derived from only one entomopathogenic bacteria, *i.e., Bacillus thuringiensis* (Bt). Major producers and users of biopesticides in the world are the United States (US), European Union (EU) and India. In the US market, more than 200 Bt products and in EU 60 Bt-based products and 60 analogous products are commercially available. According to the report of India Biopesticides Market Insights, the Indian biopesticides market is expected to reach INR 316 Billion by 2024 from INR 197 billion during

There have been growing interests in sustainable and organic agriculture in developing countries because of the availability of natural inputs, human resources and agro-climatic conditions. Disadvantages of the use of chemical insecticides and demand for organics pushed biopesticides forefront in pest management. Microbes provide certain distinct advantages over other control agents and methods. The major advantage of exploiting microorganisms for pest control is their environmental safety primarily due to the host specificity of these pathogens. Microorganisms have natural capability of causing disease at epizootic levels due to their persistence in soil and efficient transmission. Many insect pathogens are compatible with other control methods including chemical insecticides and parasitoids. The cost of development and registration of microbial insecticides are much lesser than chemical insecticides. The self-perpetuating nature of most of the pathogens in both space and time are major advantages of biopesticides. There is a long-term regulation of a pest population in nature by beneficial organisms to keep pest populations below the threshold level. Though biopesticides possess several benefits, it is not well popularized due to lack of awareness among the farmers. It is necessary to convince farmers and government agencies to adopt biopesticides as a prime component of IPM.

2018. Biopesticides market potential in India is remarkably huge as the crop losses due to pest is about USD 9259 million per annum. However, presently, the biopesticides market accounts for only 5 per cent of the pesticide industry. The Central Insecticides Board and Registration Committee (CIBRC) in India have registered about 970 microbial formulations. Further, about 200 entomopathogenic fungi-based products are available for commercial use. Among them, *Beauveria bassiana, B. brongniartii, Metarhizium anisopliae, Lecanicillium lecanii* and *Hirsutella thompsonii* are widely used against arthropods. A total of 45 products of Bt are commercially available. *B. thuringiensis* subspecies kurstaki is widely used against bollworms, loopers and other lepidopterans. *B. thuringiensis* subspecies israelensis and sphaericus are used against the mosquito. Among viruses, nuclear polyhedrosis viruses (NPV) are used against bollworms and armyworms. Rao and Manjunath (1966) demonstrated the use of first entomopathogenic nematodes *Steinernema carpocapsae* (DD-136 strain) for the control of lepidopteran pests of rice, sugarcane and apple in India. Several other indigenous strains of entomopathogenic nematode such as *Heterorhabditis indica, Steinernema carpocapsae, S. abbasi, S. thermophilum, S. riobrave* were also exploited for the management of field pests. The development of wettable formulation from *H. indica* with an extended shelf life of 12 months is commercially available in India (Kumar *et al.*, 2019).

11.3 Categories of Biopesticides

Based on the classification by Environmental Protection Agency on the type of ingredient used, biopesticides are categorized into three major classes:

1. Microbial,
2. Biochemical and
3. Plant-incorporated protectants).

11.3.1 Microbial Pesticides

Microbial pesticides are the largest group of pest-specific, broad-spectrum biopesticides. Microbial pesticides include use of microbes such as bacteria, viruses, fungi and protozoans as active ingredients for the management of insect pests. They are relatively precise for their target species. Microbial biopesticides are self-perpetuating, host-specific and environment-friendly. Among the most widely used microorganisms against insect pests is Bt. It is used to control a wide array of pests including lepidopterans, coleopterans and dipterans.

11.3.2 Plant-Incorporated Protectants (PIPs)

They are also called as genetically modified crops. Plant-incorporated protectants are pesticidal substances produced by plants and the genetic material required to produce such substances are introduced into the plants to offer resistance against pests. Pesticidal proteins separated from the bacteria or fungi are introduced into the plant and the genetically modified plants resist against specific pests. A typical example of this is use of Bt protein to develop PIP thorough the process of genetic engineering. Bt toxin is hostspecific, achieves quick mortality of the pests usually within 48 h. No harmful effect on the ecosystem and it does not harm vertebrates.

11.3.3 Biochemical Pesticides

They are also known as herbal pesticides. They are naturally occurring substances and secondary metabolites that control or inactivate pests. Most widely used biochemical pesticide is from neem and neembased formulations such as neem oil, neem seed kernel extract, neem extract concentrates from bark and leaves which are available in India. Also, essential oils from canola, tea tree, lemongrass (*Cymbopogon citrates*) and pyrethrin from *Chrysanthemum cinerariaefolium* are used as biochemical pesticides. Diatomaceous earth (DE) derived from fossilized sediments of numerous marine and freshwater siliceous organisms especially diatoms and other algae are used against an array of field pests. It has high absorption potential cause abrasion and desiccation in insect cuticle and finally results in death of insect.

Semiochemicals come under biochemical pesticides which include insect attractants, repellents, sex pheromones and kairomones. Semiochemicals are chemical signals produced by one organism usually by insects which results in behavioural change in an individual of the same or different species. The most widely used commercially available semiochemicals are insect sex pheromones. Sex pheromones are very successful to control adults by causing mating disruption. Mass trapping using sex pheromones to manage flying insects especially lepidopterans and coleopterans are very useful in IPM programs.

11.4 Biopesticides Derived from Bacteria

Bacterial pesticides are broadly classified into three categories:

1. **Obligate pathogens:** These pathogens require special conditions and media for growth reproduction and sporulation. They have a very narrow host and most of them are spore formers. They are ideal for biological pest control programs. They have good stability. For example: *Bacillus papillae* and *B. lentimorbus* cause milky disease in white grub populations.
2. **Facultative pathogens:** Facultative pathogens do not require special conditions for growth, reproduction and sporulation. They are less virulent than obligate pathogens. Two categories of facultative pathogens are crystalliferous and non-crystalliferous. Crystalliferous bacteria produce proteinaceous crystals responsible for its toxicity. These are mostly spore formers. They have a wide host range. The bacterium *B. thuringiensis* berliner is a crystalliferous spore former commercially exploited as the biocontrol agent.
3. **Potential pathogens:** Potential pathogens can easily be cultured on artificial media and have a wide host range. It can infect insects even at small doses and has a broad host range, *e.g. Pseudomonas aeruginosa* and

Serratia marcescens used against root-knot nematodes and *Phyllophaga blanchardi* used against coleopteran pests. The non-invasive and non-spore forming nature of these pathogens limits them as promising biocontrol agents.

Microbial biopesticides from bacteria successfully used in mass multiplication are *Bacillus popilliae, B. thuringiensis, Clostridium bifermentans, Pseudomonas alcaligenes, Pseudomonas aureofaciens, Saccharopolyspora spinosa, Serratia entomophila and Streptomyces avermitilis. B. thuringiensis* and *Bacillus sphaericus* are harmful to lepidopterans, dipterans, coleopterans and hymenopteran pests. The insecticidal properties of this bacterium lie during the spore formation phase. During sporulation, Bt produces specific crystal protein (Cry I-IV and Cyt) which is harmful to the target pests. Table 11.1 enlists the B. thuringiensis subspecies and its toxin used against various pests. The B. thuringiensis based pesticides commercially available to control various pests (Gelernter and Schwab, 1993) are presented in Table 11.2.

Table 11.1. List of *Bacillus thuringiensis* Subspecies and its Toxin Used against Various Pests

Bacillus thuringiensis Subspecies	*Toxin Class*	*Prototoxin Size (K Da)*	*Target Insect*
Bt subsp. Berliner	Cry I	130–140	Lepidoptera
Bt subsp. Kurstaki	Cry I	130–140	Lepidoptera
Bt subsp. kurstaki	Cry II	71	Lepidoptera, Diptera
Bt subsp. aizawai	Cry I	135	Lepidoptera
Bt subsp. tenebrionis	Cry III	66-73	Coleoptera
Bt subsp. israelensis	Cry IV	68	Diptera

Table 11.2. List of *Bacillus thuringiensis* based Pesticides Commercially Available to Control Various Pests

Bacteria Used	*Target Pests*	*Trade Name*
Bacillus popilliae	Japanese beetle grubs	Doom, Japidemic, Milky Spore Disease, Grub Attack
B. sphaericus	Mosquito larvae	Sphericide, VectoLex
B. thuringiensis subsp. *aizawai*	Lepidopterans	Certan
B. thuringiensis subsp.*israelensis*	Mosquito, blackflies and fungus gnats	Bacticide, Summit Bactimos, Bacto Power-Bti, Biodart M, Larvect50, Deltafix, Biovectra Spicbiob, VectoBac, Vectocid

Bacteria Used	*Target Pests*	*Trade Name*
B. thuringiensis subsp. *kurstaki*	Lepidopteran larvae	Abtec Btk, Agni, Bioasp, BioDart, Bactur, Biobit, Bioworm, Biolep, B.T.Killer, Caterpilin, Cezar, CID, Deflin, DiPel, Dipole, Gold Btk, Halt, JasBT, KavachBt, KrishiBio, Prasar, Lipel, Mahastra,Minchu, Neelstaki
B. thuringiensis subsp. *tenebrionis*	Colorado potato beetle, stored pests, coleopteran adults and larvae	Foil, M-one, M-track, Novardo, Trident
B. thuringiensis subsp. *Galleriae*	Lepidopteran larvae,boll-worms, Diamondback moth	Spicturin

Source: Usta (2013).

11.4.1 Mode of Action of *Bacillus thuringiensis*

B. thuringiensis produces protein crystals during spore formation, which is responsible for lysis of gut cells when consumed by susceptible insects. *B. thuringiensis* produces crystalline protein (cry and cyt). When cry protein is ingested by the larva, delta-endotoxins are activated in the gut of the insects which has alkaline pH (9.0 to 11.0). The proteins are attached to the gut receptor site, create pores in the midgut cells. Bt toxin causes paralysis of the midgut, cell lysis results in the release of gut contents into the hemocoel of the insect, disrupting the pH balance and finally death of specific target pest.

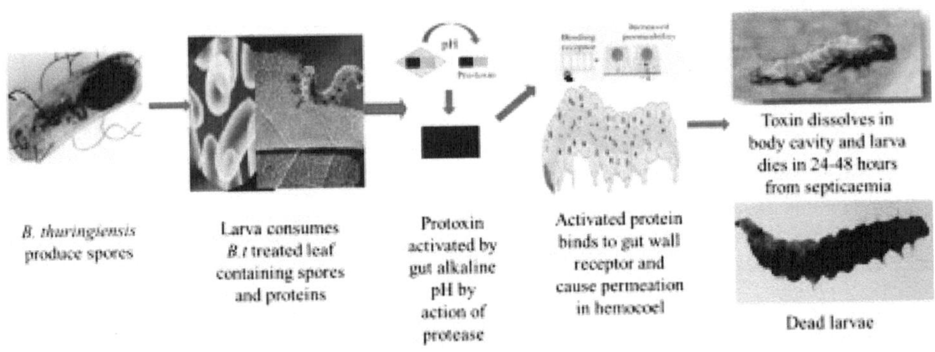

Mode of action of *B.t*

Figure 11.1. Mode of Action of Bacterial Pesticide

In contrast to Cry proteins, Cyt toxins are inserted directly into the lipid membrane of insects. Recent evidence shows that Cyt synergizes or suppresses resistance to mosquitocidal-Cry proteins by functioning as a membrane bound receptor for Cry toxin. Both Cry and Cyt toxins are pore-forming toxins.

11.4.2 Advantages of Bacterial Biopesticides

- ☆ ***Safe to use:*** Bacterial biopesticides are safe to wildlife, humans and other organisms.
- ☆ ***Specific:*** The bacterial biopesticides are specific against their target organisms. They do not affect the beneficial insects.
- ☆ ***Compatible with other methods:*** Most of the bacterial insecticides are compatible with synthetic chemical biopesticides.
- ☆ ***Non-hazardous:*** It leaves no residues. So it can be applied even during the harvesting stage of the crop.
- ☆ ***Natural establishment:*** Bacterial biopesticides are self-perpetuating. Hence it is effective even during the subsequent crop growth seasons.

11.4.3 Disadvantages of Microbial Insecticides

Microbial insecticides are toxic to only a specific species or group of insects. It may control only specific pests and the others will survive and continue to cause damage.

When predators and parasitoids are used in pest management programs use of synthetic pesticides may harm the beneficial organisms. The selection of appropriate chemical pesticides is essential.

The effectiveness of microbial insecticides is affected by factors like ultraviolet radiation, heat, *etc*. Hence, it should be applied in the morning hours only

Special formulation and storage procedures are necessary for microbial pesticides. Storage of microbial insecticides according to label directions is mandatory to maintain the virulence of pathogens.

Mass multiplication of natural enemies throughout the year is difficult thus limit the availability in the market. The product registration process and less availability limit the use of biopesticides compared to chemical pesticides.

Effectiveness of biopesticides is mainly depending upon the external environmental conditions such as temperature, relative humidity and rainfall. Weather conditions should be suitable for the multiplication and survival of biocontrol agents in the field.

11.5 Viruses as Biopesticides

Baculoviruses (BV) are widely used as viral biopesticides. They are highly pathogenic for insects and other arthropods. BV is tiny double-stranded DNA viruses and parasitically replicating microscopic elements.The genus baculoviruses containthreesubgroups:

- ☆ Nuclear polyhedrosis viruses (NPVs),
- ☆ Cytoplasmic polyhedrosis viruses (CPVs) and
- ☆ Granulosis viruses (GVs).

They are considered as inclusion viruses which differ in the number of occlusion body and structure of the protective protein coat. NPV and CPV produce polyhedral bodies which are comprised of numerous virus particles. GV produces granular occlusion bodies which contain just one virus particle.

The distinct mechanisms of NPV, CPV and GV virus uncoating occur among the baculoviruses. NPVs uncoat within the nucleus, CPVs uncoat in the cytoplasm and GVs uncoat within the nuclear pore complex to establish infection in the host. The occlusion bodies help the virus to survive outside the host (Cory 2000).

11.5.1 Mode of Action of Viruses

The mode of action of the viruses is like bacteria and needs to be ingested by the insect larvae to initiate infection. Immediately after the ingestion, virus enters the insect body through the intestinal epithelium and causes systemic infection in the host cell.

NPV passes through intestinal epithelium to hemocoel, fat body and other tissue cells, disintegrates the integrity of the tissues and liquefies the cadavers. In case of GV, the infection is limited to insect midgut.

11.5.2 Steps Involved in the Preparation of NPV and CPV

Usually third to fourth-instar larvae of *Helicoverpa armigera* are infected with viral food. The definitive phase of viral disease occurs over 5–10 days. Once the complete infection of the virus in the larvae is completed, before death, putrefying infected larvae climb higher in the plant canopy and start releasing billions of polyhedra which aid in the dissemination of virus particles from the cadavers to the lower parts of the canopy.

This behaviour helps in the spread of the virus to cause infection in healthy larvae. In commercial production, larvae are being harvested from the field or reared in the laboratory and allowed to feed on the contaminated diet. Before the inoculation of the virus, larvae are kept hungry for 24 h.

Within three to four days larvae get infected with the virus and hang upside down in tubes. Infected larvae are collected and kept in water for purification. Purified larvae are crushed using mortar and pestle with sterile chilled distilled water in a ratio of 1:2.5 (w/v). Homogenate filtered with double layer muslin cloth and repeatedly washed with sterile distilled water in the ratio of 1:7.5 to 12.5 (w/v) for the original weight of cadaver larva.

The filtrate is centrifuged to separate polyhedra from the debris. The pellets are repeatedly washed and centrifuged three times with sterile distilled water and finally collected purified polyhedra in the suspension are counted through haemocytometer. The dose of virus is fixed as one larval equivalent (LE) that is the effective lethal concentration (LC50) of NPV is $1x10^9$./mL for *H. armigera* larvae. For gram, pigeon pea and cotton HNPV at 250–300 LE, 500 LE and 250 LE/ha should be used.

Viruses are very host-specific and can cause significant reduction in host populations. Examples of some commercially available viruses include Helicoverpa zea single-enveloped Nuclear polyhedrosis virus (HzSNVP), Spodoptera exigua multi-enveloped nuclear polyhedrosis virus (SeMNPV) and Cydia pomonella granulovirus (CpGV). Table 11.3 shows the list of commercially available viral biopesticides.

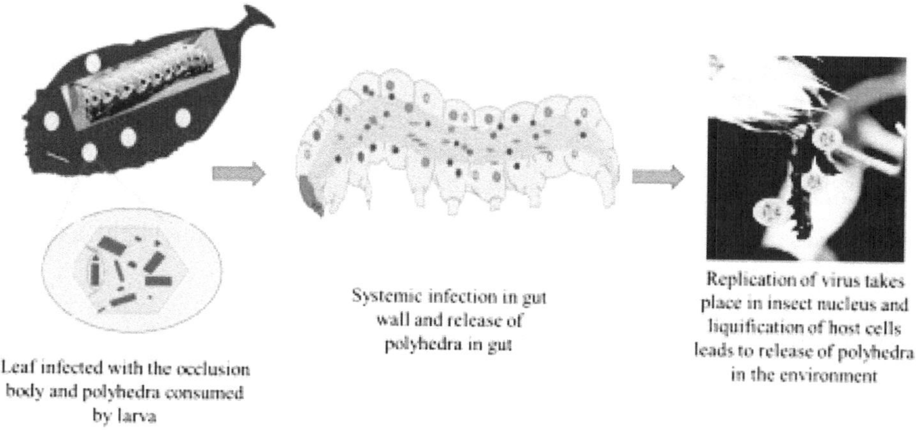

Figure 11.2. Mode of Action of Virus.

Table 11.3. List of Commercially available Viral Biopesticides

Target Pest	*Host*	*Products available*
NPV of *H. armigera* (HaNPV)	*Helicoverpa zea, H. armigera*	Biokill-H, BioVirus-H, Heli-Cide, Heliokill, Helimar-NPV, Helivax, Jas Viro–H, Helicop, Heligard, Somstar-Ha
NPV of *S. litura* (SlNPV)	*Spodoptera litura, S. exigua*	BioVirus-S, Jas Viro–S, Spodo-Cide, Spodopterin, Somsta-SL
Nuclear polyhedrosis for Gypsy moth	Gypsy moth caterpillars	Gypchek virus
Tussock moth NPV	Tussock moth caterpillars	TM Biocontrol-1
Pine sawfly NPV	Larvae of pine sawfly	Neochek-S
Granulosis virus for Codling moth (GV)	Codling moth caterpillars	Madex, Carpovirusine, CYD-X

Source: Usta (2013).

11.5.3 Advantages of Viral Biopesticides

Viruses are safe to humans and other non-target organisms.

The target organisms will not develop resistance against viral biopesticides.

Viruses can be integrated with other methods of insect control including chemical pesticides.

They are self-perpetuating and keep the pest population under check.

No secondary pest outbreaks.

No pesticide residues.

No pre-harvest interval is required.

11.5.4 Disadvantages of Viral Biopesticides

It can be used for the management of multiple pests as that of chemical pesticides.

A long period of lethal infection is required for successful control of target pests.

Viruses get inactivated by environmental factors like ultraviolet light, extreme temperature, *etc.*

11.6 Fungi as Biopesticides

Fungi-based biopesticides are most versatile since it has a wide host range to the field and stored pests. They are safe and highly specific. Fungal products are developed using saprophytic or parasitic fungi, which grows inside an insect's body and feed on its internal tissues.

The most common commercial fungal biopesticides available in the market are from *Beauveria bassiana, Metarhizium anisopliae, Nomuraea rileyi, Trichoderma viride, Paecilomyces farinosus and Verticillium lecanii.* These fungi are either obligate or facultative, has symbiotic or antagonestic relationship with insects for their survival. The pathogenic action of fungi depends on the contact and subsequent infection to the target pests.

Commercial formulations of B. bassiana and M. anisopliae are effective against aphids, beetles and their grubs, grasshoppers, leaf and plant hoppers, pod borer, cutworms and other lepidopteran and coleopteran pests (Table 11.4).

11.6.1 Mode of Action of Fungi-Based Biopesticides

Fungi act by penetration of the mycelium into the cuticle of the insects into the hemocoel causing the death of the attacked individuals.

- ✰ The fungal spore called conidia get adhere to the host cuticle by hydrophobic attachment.
- ✰ Spore germination will take place through the filamentous growth of conidia called germ tube or by appressorium.
- ✰ It penetrates the cuticle through hydrolytic enzymes produced by them. Also, organic acids, secondary metabolites, toxins and detoxifying antimicrobial compounds are secreted by fungi.

Table 11.4. List of Commercially available Fungi-based Biopesticides

Fungi	*Host Range*	*Products*
Beauveria *Bassiana*	Effective against a variety of insects such as rickets, white grubs, fire ants, flea beetles, plant bugs, grasshoppers, thrips, mites, mosquito larvae and aphids, fungus gnats, whiteflies and to stored pest	BotaniGard, Mycotrol, Naturalis, Racer BB, Ostrinil, Brocari l, Mycotrol, Mycotrol-0 BotaniGard, Boverin, Naturalis-L, Naturalis- H and G Naturalis-T and O
Metarhizium anisopliae	Effective against a wide range of pests. Black vine weevil, locusts and grasshoppers	Bio-Blast, Bio-Path, Green Muscle, Met52 Metaquino Pacer MA Tick- Ex
Trichoderma viride	Effective against Rot disease	EcosomTV, Tricon, Trieco
Trichoderma harzianum	Effective against a variety of soil pathogens	Rootshield, BioTrek, Supresivit

Source: Usta (2013).

- ☆ Fungal spores penetrate from the cuticle of the insects into the hemocoel leads to the death of insects.
- ☆ Further, the internal mycelium grows outward sporulate and emerges on the outer surfaces of cadavers of insects.

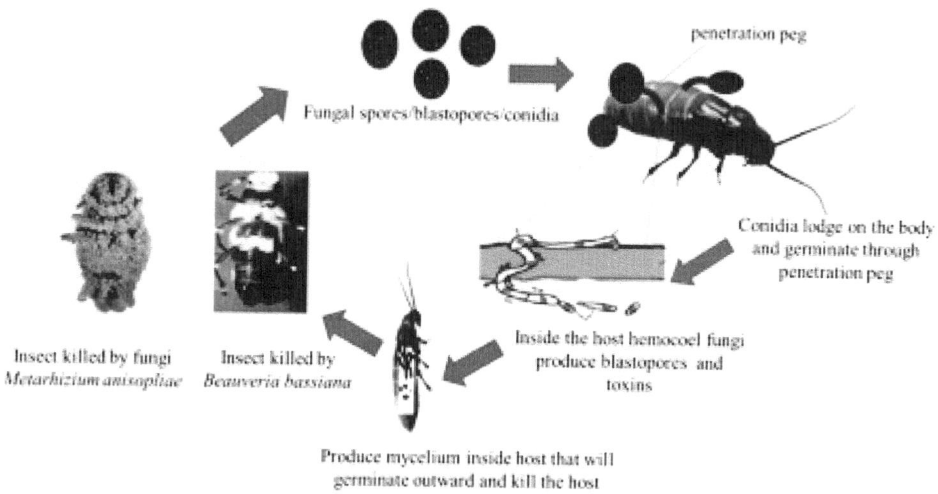

Figure 11.3. Mode of Action of Fungi-Based Biopesticides.

11.6.2 Advantages of Fungi-Based Biopesticides

Wide host range: As compared to bacteria and viruses, fungi have a wide host range and can infect field, stored and soil pests.

Ease of production: Commercially important fungi such as Beauveria, Metarhizium, Lecanicillium and Isaria are relatively easy to mass produce with low substrate requirements.

Effective and highly specific.

Eco-friendly.

11.6.3 Disadvantages of Fungi-Based Biopesticides

Ideal environmental conditions: Fungi are highly sensitive to environmental conditions. Humidity over 80 per cent and above is required for spore germination and penetration on the surface of arthropods cuticle.

Temperature and exposure to UV radiation affect the survival of fungi.

Production and stabilization of fragile conidia or durable resting phase are difficult.

It is costlier than other microbial methods

11.7 Entomopathogenic Nematodes (EPN) as Biopesticides

Entomopathogenic nematodes are soil-dwelling, soft-bodied, un-segmented roundworms that are naturally present in water film near soil particles. Nematodes are obligate or sometimes facultative parasites that locate the host insects in response to carbon dioxide, vibration and other chemical indications.

Commercially two families, namely Steinernematidae and Heterorhabditidae, have been effectively identified and used as bionematicides in pest control programs. *S. carpocapsae, S. thermophilum, H. bacteriophora, H. indica* products are registered and sold in India.

11.7.1 Mode of Action of EPN

The non feeding infective third stage juveniles (0.4–1.5 mm in length) enter into an insect through natural openings, such as mouth, anus or breathing holes.

Nematodes enter insect body cavity and release their symbiotic bacteria into the intestine of the host. Toxins produced by the bacteria kill the insect in a week.

It feeds on the cadaver and liquefies its host. Nematodes reproduce and generate to three-generation descendants on host cadaver.

Infective juvenile nematodes leave the dead insect and seek for a new host. Under optimal conditions, nematode-infected pest stages should be present 5–7 days after application. Insects killed by Steinernematidae become brown or tan; ones killed by Heterorhabditidae turn red. Dead insects, however, are not always visible. Table 11.5 represents the list of commercially available EPN based biopesticides.

11.7.2 Advantages of EPN

They are considered safe to humans, plants and animals and relatively safe to the environment. It does not require personal protective equipment, safety masks and re-entry intervals. It leaves no residues.

It is effective against field pests such as cranberry girdler, root weevil, black vine weevil, webworms, cutworms, armyworms and wood-borers.

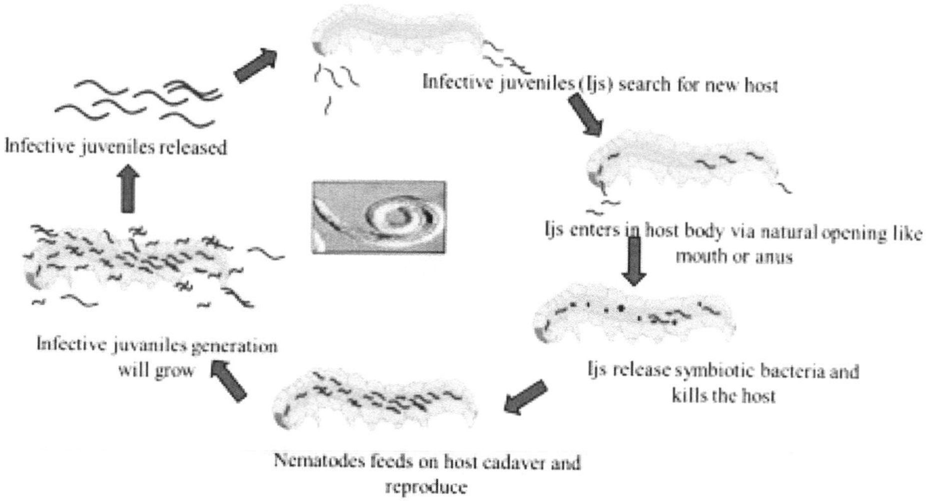

Figure 11.4. Mode of Action of Nematodes.

Table 11.5 List of commercially available EPN based biopesticides

Entomopathogenic Nematodes	*Host Range*	*Commercial Products*
H. bacteriophora	Effective against root weevils, cutworms, fleas, borers and fungal gnats. Effective against black vine weevil populations by 56–100 per cent	BioSafe, Larvanem, Nemaplant, NemaShield-HB, Nematop, Nematech-H NemaTrident-H, NemaTrident-C, Nema-green,Optinem-H
H. indica	*H. armigera, Conogethes punctiferalis, Athalia proxima*	Soldier, Nema power, GrubTerminator, Grubcure, Calterm, Aarmour
H. downesi	Black vine weevil	NemaTrident-CT
Steinernema feltiae, S. riobravis	Effective against black vine weevils, strawberry root weevils, cutworms, cranberry girdler and termites	Biosafe, Ecomask, Hortscan, Guardian, Millenium, Nematac C, No Flea, Savior WG, Scanmask, Termask, Vector
S. carpocapsae	Borer beetles, caterpillars, crane fly, moth larvae, coconut rhinoceros beetle	Capsanem, Carpocapsae-system, Exhibitline SC, Optinem-C, NemaGard, Nemastar, NemaTrident-T, NemaRed, Nemasys-C, Palma-life

Source: Koul (2011), Gupta *et al.* (2010), Ruiu (2018).

The soil around the roots of cranberries should be moist and humid; protection from harmful UV light and temperatures is often suitable for the survival of nematodes.

Nematodes can withstand short-term exposure of (2–24 h) to many chemical and biological insecticides, fungicides, herbicides, fertilizers and growth regulators.

11.7.3 Disadvantages of EPN

Nematodes are sensitive to environmental conditions such as UV and high temperatures; conditions that may be hard to control.

Compared to traditional chemical insecticides, nematodes are generally very expensive. In a natural system, the number of insect hosts and nematodes is in balance. If the natural balance is overloaded with nematodes, they will kill the insects but will then die off themselves until insect populations start to build again. Hence re-application of nematode is essential.

Since nematodes cannot live above the soil, they cannot infect insects that live above soil such as foliar-inhabiting insects. Nematode formulations have a short shelf life.

It should be applied early in the morning or in the evening to protect from UV radiation.

Pre- and post-irrigation is required for moistening the soil and washing the nematodes into the soil.

11.8 Protozoans as Bbiopesticides

Protozoans are also referred to as microsporidians. Microsporidia are ubiquitous, obligate intracellular parasites have the potential to attack lepidopteran and orthopteran insects hence can be used as a component of integrated pest management program, *e.g. Nosema* sp. and *Vairimorpha* sp.

Although, they are pest-specific, slow-acting, induce chronic and debilitating effects on targets, the use of protozoa as biopesticides is not very successful compared to other organisms such as bacteria, viruses and fungi. Table 11.6 shows the protozoa based successful commercial biopesticides.

11.8.1 Mode of Action of Protozoans

- Microsporidia must be eaten by insects to get infected.
- Microsporidian infects European corn borer, *Ostrinia nubilalis.*
- The spores germinate in the midgut region and the sporoplasm is injected into the midgut cells.
- The spores spread to different tissues and organs, where they multiply resulting in tissue breakdown and septicaemia.

11.9 Natural Enemies of Pests as Bioncontrol Agents

Natural enemies of pests such as parasitoids and predators are widely used against the field and stored pests. Insect parasitoids have an immature stage of life that develops on or within a single insect host and finally kills the host. Parasitoids lay their egg on or in the body of insect host and the developing larvae utilize host body fluids and organs as food, kill the host as it pupates or emerges as an adult. Most important insect parasitoids are from the families of Ichneumonidae and branconidae. Wasps or a fly, which attacks the caterpillar, has a wide host range including aphids. Tachinid flies parasitize a wide range of pests such as caterpillars, larvae and adults of lepidopteran pests and beetles.

For successful biocontrol program, parasitoids used should be host-specific, reproductive and should adapt to wide environmental conditions. Predators are free-living, large in size, feed on the host and destroy them during their life period. Predators such as *Brumoides* sp., *Chrysoperla* sp., dragonflies and damselflies are effective against various crop pests (Strand and Obrycki, 1996) (Figure 11.5 and Table 11.6). Table 11.7 shows the list of promising biocontrol agents used against crop pests.

Table 11.6 Protozoa based successful commercial biopesticides

Protozoan	*Host Range*	*Products*
Nosema locustae	European corn borer caterpillars, grasshoppers and Mormon crickets	NOLO Bait, Grasshopper attack

Source: Usta (2013).

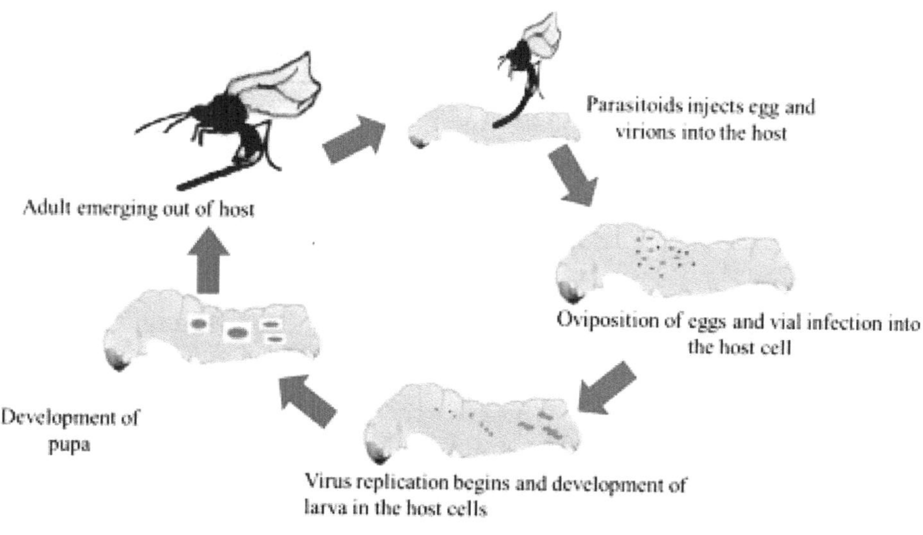

Figure 11.5. Mode of Action of Parasitoids.

Table 11.7. List of Promising Biocontrol Agents used against Crop Pests

Crop	*Pest*	*Promising Bioagent*	*Stage of Pest*	*Kind of Biocontrol Agent*
Sugar-cane	Leafhopper, *Pyrilla perpusilla* Walker	*Epiricania melanoleuca*	Nymph and adult	Parasitoid
		Ooencyrtus pyrillae Mani	Eggs	Parasitoid
	Internode borer, *Chilo sacchariphagus indicus*	*Cotesia flavipes*	Larva	Parasitoid
		Rhaconotus scirpophagae Wilk	Larva	Fungal pathogen parasitoid
		Trichogramma chilonis Ishii	Larva	Parasitoid
	Gurdaspur borer, *Acigona steniellus*	*Cotesia flavipes*	Larva	Parasitoid
		Xanthopimpla stemmator Thunberg	Pupa	Parasitoid
	Maize borer, *Chilo partellus*	*Allorhogaspyralophagus* Marsh	Larva	Parasitoid
	Root borer, *Emmalocera Depressella swinhoe*	*Goniozus indicus* Ashmead	Larva	Parasitoid
		Elasmus Zentri ferriere	Larva	Parasitoid
	Shoot borer, *Chilo infuscatellus* Snellen	*Telenomus beneficiens* Zhnt.	Egg	Parasitoid
		Tetrastichus schoenobii Ferr.	Egg	Parasitoid
	Stalk borer, *Chilo auricillus*, Dudgeon	*Beauveria bassiana*	Larva	Fungal pathogen
		Goniozus indicus ashmead	Larva	Parasitoid
		Sturmiopsis inferens Tns.	Larva	Parasitoid
		Trichogramma chilonis	Egg	Parasitoid
	Top borer *Scirpophaga excerptalis* Walker	*Cotesia flavipes*	Larva	Parasitoid
		Fusarium subglutinans	Larva	Fungal pathogen
		Goniozus indicus Ashmead	Larva	Parasitoid
		Isotima javensis	Larva	Parasitoid
		Sturmiopsis inferens Tns.	Larva	Parasitoid
	White grub, *Holotrichia consanguinea*	*Bacillus papillae* Dutky	Grub	Bacterial pathogen

11.9.1 Advantages of Parasitoids in Biological Pest Management

- Specialized and host-specific.
- Parasitoids are efficient host searchers, can find their hosts even at low pest densities and bring infestation level below economic thresholds.
- Parasitoids reduce the number of pests surviving to the next generation.
- They are compatible with other biological control agents (predators).
- Some parasitoids affect feeding behaviour, *e.g.* parasitized larvae eat less than healthy ones, and are smaller than the host.

11.9.2 Disadvantages of Parasitoids in Biological Pest Management

- ✰ Highly susceptible to insecticides, especially adults.
- ✰ Adults require an alternative food source such as pollen or nectar.
- ✰ Immature parasitoids, especially if protected within the egg of their host or in their own cocoon, may tolerate pesticides better than adults, but immature parasitoids will usually die if their host is killed.
- ✰ To improve the abundance and diversity of beneficial insects especially parasitoids and predators it is necessary to consider native vegetation on the farm as a part of a pest management strategy.
- ✰ Sometimes multiple parasitoids may emerge from one pest.

11.9.3 Advantages of Predators in Biological Pest Management

- ✰ Bigger in size, prey many hosts, can reduce the pest population potentially.
- ✰ Not much affected by external environmental conditions and pesticides compared to parasitoids.

11.9.4 Disadvantages of Predators in Biological Pest Management

- ✰ Non-host specificity.
- ✰ Mass multiplication is difficult.
- ✰ Cannibalism is common in many predators.

11.10 Biochemical Pesticides

Phytochemicals or herbal pesticides are chemicals derived from plant origin. Each phytochemical possesses specific structure, performs various functions such as protection, growth acceleration and reproduction in plants. Phytochemicals are found in various parts of the plants like fruits, vegetables, grains, pulses, nuts, seeds, bark, *etc.* Since it has several bioactive compounds that contain antiparasitic, bactericidal, fungicidal, viricidal and insecticidal properties, it can be considered as a potential alternative for inorganic pesticides.

These are secondary metabolites produced by plants by nature such as:

- ✰ Terpenes contain phytovolatiles, glycosides,
- ✰ Sterols contain phenolic compounds such as phenolic acids, lignin, tannins and alkaloids. Secondary metabolites play an important role in the plant defence system against insects. It acts as a toxicant, insect growth regulator, repellents and antifeedant. Phytochemicals from the plants can be extracted using solvent extraction, microwave-assisted extraction, ultrasonic assisted extraction based on the presence of a group of phytochemicals present in it. At present only a few phytochemicals are used as biopesticides and most of them are neem-based (Table 11.8).

These are mainly 3 per cent neem oil or 5 per cent neem seed kernel extract concentrates which can be applied at 25 kg/ha against major field pests.

Table 11.8. List of Commercially Available Biochemical Pesticides

Plant Source	*Target Pest*	*Commercial Name/Compounds*
Azadirachta spp., *Nicotiana* spp.	*Aphis craccivora*, Bollworms, Aphids, Jassids, Thrips, Whitefly, Leaf folder, Pod borer, Fruit borer, Leafhopper, Diamondback moth	Essential oils, nicotine, azadirachtin, salanin, nimbin, (leaf, seed, kernel) nimbecidin, bionimbecidin
Chrysanthemum cinerariaefolium *Lonchocarpus* spp. *Ryania* spp.	Crawling and flying insects such as cockroaches, ants, mosquitoes, termites	Pyrethrium, Rotenone Ryanodine
Artemisia annua	*Helicoverpa armigera*	Essential oils from leaf and seed
Vinca rosea	*Helicoverpa armigera*	Essential oils from leaf and seed
Ocimum basilicum *Salvia Officinalis*	*Aphis craccivora* Koch, *Agrotis ipsilon*	Essential oils form aerial parts leaves
Citrus sp.	Fleas, aphids, mites, paper wasp, house cricket	D-limonene linalool
Schoenocaulon officinale	Bugs, blister beetles fly, caterpillars, potato leafhopper	Sabadilla dust
Adenium obesum	Cotton pests	Chacals Baobab (Senegal)

11.10.1 Mode of Action

The mechanism of action of herbal pesticides is either by absorption by insect cuticle, inhalation or by ingestion thus affects the physiology, biochemical pathway or nervous system of the insects. It affects the respiration of insects by blocking spiracles leading to suffocation. At the cellular level, it will interfere with the signal transduction, disturbing ion channel and ion pump of insect membrane.

It also affects the neuroendocrine system of insects and juvenile hormonal imbalance resulting in malformation of insects. The decrease in adult mating time, reduction in ovulation, reduced fecundity and egg fertility are other effects.

Monoterpenes present in the essential oil act as neurotoxicant which acts on the acetylcholinesterase enzyme involved in nerve impulse transmission in insects causing paralysis and death of the insects. It also interferes with DNA, RNA and protein synthesis (Figure 11.6).

11.10.2 Semiochemicals

Semiochemicals are chemical substances produced by the organisms that enable and mediate communication between the organisms. Basically there are two types of semiochemicals, namely intraspecific semiochemicals (pheromones) and interspecific semiochemicals (allelochemicals).

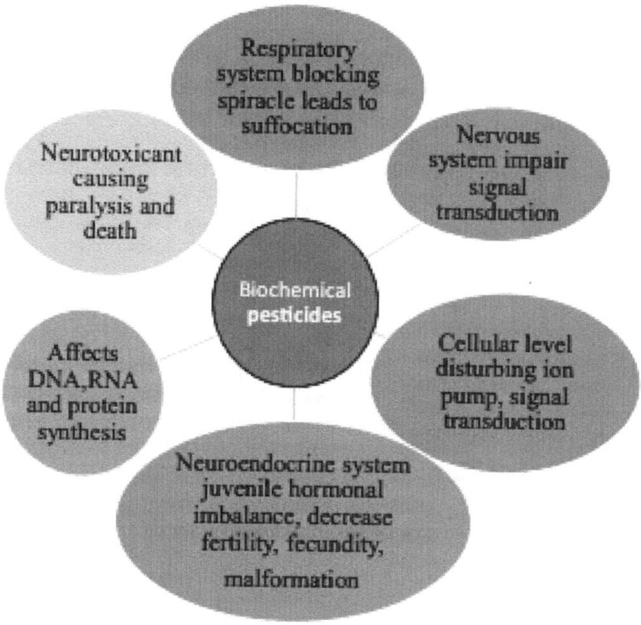

Figure 11.6. Mode of Action of Biochemical Pesticides.

Pheromones are volatile chemical substances produced by an organism so as to elicit a reaction or response from the receiving individual (of same species). The releaser pheromones cause behaviour change in the recipient. The releaser pheromones include sex pheromones, aggregation pheromones, alarm pheromones and trail pheromones. Aggregation pheromones are released by insects as an indication of food source available.

Alarm pheromones are secreted by insects to show the presence of a predator and cause movement of insects away from the source. Sex pheromones are released normally by one sex only and they elicit responses in the other sex. These pheromones facilitate in mating.

Synthetic analogues of sex pheromones are used in pest management. In fact, sex pheromones are used for pest monitoring, mating disruption and mass trapping.

11.10.3 Advantages of Biochemical Pesticides

- ☆ Safe for natural enemies, environment and human beings.
- ☆ Lesser chance to develop resistance to these pesticides.
- ☆ Economical, biodegradable, renewable and user friendly.
- ☆ Active component degrade rapidly thus it is more acceptable.

- They are stable and can be stored for extended period.
- They are insect specific especially in case of semiochemicals.

11.10.4 Disadvantages of Biochemical Pesticides

- More quantity of pesticide required due to its crude formulations.
- Mode of action is slow.
- Higher specificity and require an exact identification of the pest/pathogen.
- Variable efficacy.

11.11 Plant-Incorporated Protectants

Plants with the toxin-producing genes to fight against the pest are called plant incorporated protectants or genetically modified or genetically engineered plants. These PIPs help the plants to resist bacteria, viruses or fungi. When PIPs are used against target insect, it is known as insect-resistant crops.

Some of the examples for PIPs are soybean, brinjal, cotton, corn, potato and tomato. Transgenic plants are developed by transferring the target gene either by Agrobacterium-mediated transfer or by gene gun or ballistic methods (Figure 11.7). These methods were used in wheat, maize, rice and corn (Table 11.9). Plant-incorporated protectant have several advantages like it will give more yield with lesser amount of pesticide use. Thus, they are cost effective and protect neighbouring non-Bt crop also.

Table 11.9. List of Plant-Incorporated Protectants against Crop Pests

Plant	*Gene Incorporated*	*Resistance to Insect*	*Commercial Name*
Cotton	Cry 1Ac toxins from *Bacillus thuringiensis*	Cotton and tobacco bollworm	Bollgard
Maize	Cry 1Ab,1F,9c toxins from *Bacillus thuringiensis*	European corn borer	Yieldgard, Knockout, Herculex I, Starlink, Maximizer
Rice	Cry1C from *Bacillus thuringiensis*	Leaf folders (*Cnaphalocrocis medinalis*) and stem borers	
Tobacco	QB protein of photosystem II from mutant Amaranthus	Atrazine resistant	-
Brinjal	Cry1Ac from the soil	Shoot and fruit borer, *Leucinodes orbonalis*	*Bt* Brinjal
Tomato	Cry1Ab gene of *Bacillus thuringiensis*	Second instar larvae of *H. armigera* and *S. litura*	*Bt* tomato

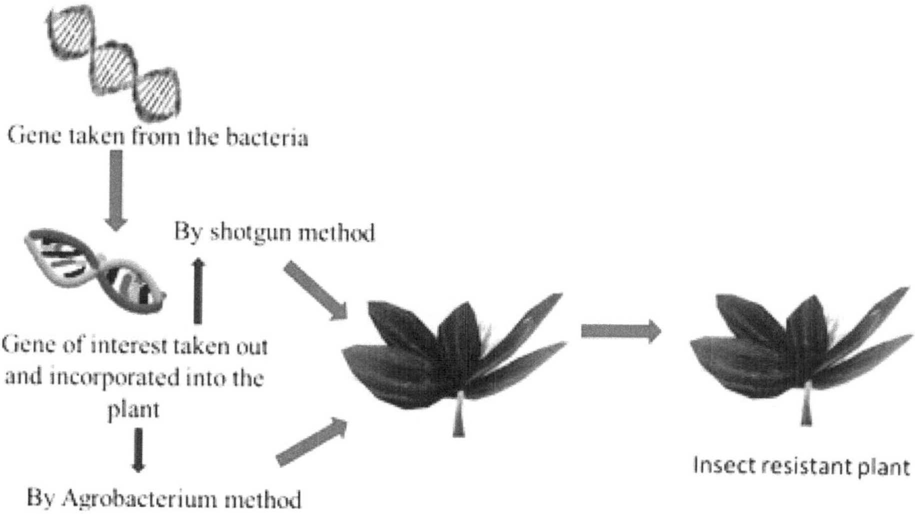

Figure 11.7. Technique Producing Plant-Incorporated Protectants.

Some potential risks are also associated with the plant-incorporated protectants, such as it affects non-target pest, risk to human health and environment, possibility of PIP gene to spread to the other plants and chances for development of pesticide or insect resistance.

11.12 Biopesticides Formulations

In most cases, biopesticides are formulated in the same way as that of synthetic pesticides. They are more convenient to use by the farmer and same equipment can be used for an application. Since it is made from living organisms, it is a prerequisite to maintain its viability during formulation and storage.

Biopesticide formulations are available as active ingredients (AI) and inert or inactive ingredients. The active ingredient(s) is part of the formulation designed to control the target pest. The function of inert ingredients is to enhance application and effectiveness of the active ingredient(s).

A pesticide formulation is nothing but a combination of active and inactive ingredients. The organism used in the biopesticides formulation must revive from their dormant state to be active during application and post-application phase (Boyetchko *et al.,* 1999).

Biopesticides are divided into two categories based on formulations.

1. Dry (dust powders, granules, seed dressers, wettable powders and wettable dispersible powder)
2. Liquid (Emulsion, suspension, emulsifiable concentrate, ultra-low volume liquids)

11.12.1 Dry Powders

Dry formulation is used for direct application as dustable powders in which only 10 per cent of active ingredients are mixed with a finely ground mineral powder such as clay or talc. The inactive ingredients act as UV protectants and they increase the adsorption of biopesticides.

The active ingredient in granule formulation ranges from the 2–20 per cent which is coated or absorbed into granules made up of silica, starch, clay, groundnut plant residues, *etc*. These formulations are applied near to soil. When sufficient soil moisture is attained, it releases its active component. It can control insects, weeds and nematodes. In India, Bt var. israelensis Serotype H-14, Strain VCRC B-17 is available in the form of slow-release granules.

Seed dressing/coating formulations are also available in powder form. They are mixed with seed before planting. Wettable powders (WP) are finely ground powders obtained by mixing active ingredients with dispersing agents, surfactants, melting agents, *etc.* and can be applied after mixing with water. Wettable powders have good storage, stability and miscibility with water.

In India, Bt var. kurstaki and Bt var. israelensis (5 and 2.5 per cent WP); *Trichoderma viride* 1 per cent and 1.5 per cent WP; *Pseudomonas fluorescens* 1.25 per cent WP; Entomogenous fungi *Beauveria bassiana* are available in wettable powder formulations. Water dispersible granules (WDG) are dust-free formulations that can be suspended with water to enhance storage stability.

11.12.2 Liquid Formulations

Active ingredients of liquid formulations are mixed with water/aqueous suspension (AS) or with oil, kerosene or petroleum-based solvents as a carrier. *B. thuringiensis* var. kurstaki is available in 2.5 and 3.5 per cent AS. NPV of *H. armigera* is used in 0.43 per cent (AS), and NPV of S. litura in 0.50 per cent (AS).

Various types of liquid formulations are available in the market including emulsifiable concentrate. An emulsion is a mixture of homogenous emulsion with active ingredient particles suspended into it.

In emulsion, active ingredient is in liquid form and is dissolved in the petroleum based solvents. To improve the stability of the emulsion, it is necessary to choose an appropriate emulsifier. In suspension concentrate (SC), which is also known as a flowable concentrate, the active solid ingredient is dissolved in water. Agitation during formulation is being done for the even distribution.

It is safer for the environment and user. *B. thuringiensis* var. galleriae 1.3 per cent, *B. thuringiensis* var. sphaericus B-101 and *Bacillus subtilis* are available in 2 per cent suspension concentrate. Entomopathogenic fungi-based SC formulation of *Beauveria bassiana* is available as 1.15 per cent, 10 per cent SC. Ultra-low volume formulation (ULV) contains 100 per cent active ingredient. It is used for mosquito control and has agriculture and forestry application. Newer formulation, capsule suspension is developed in which active ingredients are encapsulated in starch/

cellulose or other polymers that can be dissolved in water before use. This type of formulation is used to protect the bioagents from harsh unfavourable conditions.

Liquid aerosol is also used against field pests where the active ingredients are kept in the form of liquid or the gaseous phase (in case of fumigants) in a pressurized container. Thus, by pushing nozzle, very fine-droplets of liquid or gas can be applied to the target surface.

11.12.3 Compatibility of Biopesticides

It is the ability to mix various pesticides without affecting physical and chemical nature that results in increased effectiveness and reduced phytotoxicity. Compatibility is an important aspect to be considered while using combination of biopesticides.

One of the limitations with biological pesticides is that it has a very narrow host range and is effective at controlled environmental conditions. The sole use of biopesticides is not sufficient to control plant-feeding insect or mite populations. Thus, to increase the effectiveness of biopesticides, appropriate compatible conventional pesticides can be mixed to increase the synergism.

Also "biorational" or "reduced risk" agents such as essential oil, diatomaceous earth, insect growth regulator and related compounds can be used to enhance the toxicity.

Microbial pesticide formulation based on Bt subspecies kurstaki and Bt subspecies aizawai were found to be effective against a wide range of lepidopteran pests. Bt subspecies kurstaki (0.2 per cent) in combination with 5 per cent neem oil or citronella oil is effective against *H. armigera*.

Use of HaNPV 250 LE/ha in combination with endosulfan 0.035 per cent and Fenvalerate (0.005 per cent) +NPV; monocrotophos (0.035 per cent) + NPV; NPV + B.t and HNPV + NSKE (2.5 per cent) were found to be effective against larvae of *H. armigera*, *S. litura* and *L. trifolii*. *B. bassiana* and *M. anisopliae* fungi based pesticides in combination with deltamethrin or dimethoate 0.015 per cent and acetamiprid 0.004 per cent were effective on various field pests.

Combination of *B. bassiana* and M. anisopliae + insecticides was successful in controlling yellow fever mosquito, *Aedes aegypti*, coleopterans and stored pests.

Synergistic effect of fungal biopesticides and diatomaceous earth in the management of several crops and stored pests and B. bassiana + imidacloprid and DE were effective against soil insects such as the southern mole cricket, *Scapteriscus borellii*.

11.13 Factors Influencing the Success of Biocontrol Agent

There are several factors both extrinsic and intrinsic factors that challenge the effectiveness of biopesticides on a large scale. Since they are from biological agents, they are affected by external factors such as sunlight and UV radiation.

Ultraviolet light is extremely destructive and results in the photodecomposition of biopesticides. Temperature plays a significant role and extreme temperature results in reduced efficacy. Maintaining the optimal temperature in the field is difficult. However, for its survival and multiplication, the environmental conditions must be favourable to them.

Most of the fungi-based formulations are temperature and humidity sensitive and conidia will not grow above or below 80 per cent relative humidity. In case of NPV and other viruses, the suspension sprayed when adhere to the soil surface or leaf which is not consumed by the larvae does not have any effect on the pests.

Hydrolysis in combination with photodecomposition is often involved in the breakdown of various pesticide active ingredients. The faster the hydrolysis, the less time the pesticide is available in the environment. Leaching also affects the biopesticides.

If an acidic pesticide is applied to an alkaline surface or vice versa it may breakdown very rapidly. Thus, the soil pH and soil biota are important which influence the effectiveness of biopesticides.

Rain during the application of biopesticides wash off from the plant and reduces its effectiveness.

Some biopesticides are incompatible with chemical pesticides and thus reduce the activity of biopesticides.

The compatibility of biopesticides with predators and parasitoids is also to be considered while releasing biocontrol agents.

Antagonistic effects of phytochemicals on biopesticides are also to be considered.

Appendices

Appendix 1: Organic Crop Production Standards

1. Crop Production Plan

The producer is required to develop an organic crop production plan including:

i. Details of crops to be grown, field wise.
ii. Description of practices and procedures.
iii. List of inputs used.
iv. Source of organic planting material (seeds and seedlings).
v. Descriptions of monitoring practices and procedures.
vi. Description of management practices.
vii. Description of record keeping system.

2. Conversion Requirements and Duration

For a farm and its crop production products to be certified organic under these rules it is mandatory that the farm and entire farming operations of organic production unit has under gone a period of conversion, complying with all the standard requirements under these rules for following period:

i. 36 months for perennial plants
ii 24 months for plants/crops other then perennials

A reduction in conversion period up to 12 months can be granted if de-facto requirements under these standards are being met from last several years and this can be verified from various sources and necessary documentary evidences are available.

3. Landscape

Organic farming shall contribute beneficially to the ecosystem. The certification programme shall set standards/procedures for a minimum percentage of the farm area to facilitate biodiversity and nature conservation.

4. Choice of Crops and Varieties

All seeds and plant material shall be certified organic. Species and varieties cultivated shall be adapted to the soil and climatic conditions and be resistant to pests and diseases. When certified organic seed and plant materials are not available, chemically untreated conventional seed and plant material shall be used. The use of genetically engineered seeds, transgenic plants or plant material is prohibited.

5. Diversity in Crop Production and Management Plan

The basis for crop production in organic farming shall take into consideration the structure and fertility of the soil and the surrounding ecosystem, with a view to minimizing nutrient losses. Where appropriate, the organic farms shall be required to maintain sufficient diversity in a manner that takes into account pressure from insects, weeds, diseases and other pests, while maintaining or increasing soil organic matter, fertility, microbial activity and general soil health.

6. Nutrient Management

Sufficient quantities of biodegradable material of microbial, plant or animal origin produced on organic farms shall form the basis of the nutrient management. Non synthetic mineral fertilisers and brought-in bio fertilisers (biological origin) shall be regarded as supplementary. Mineral fertilizers shall only be used in a supplementary role to carbon based materials. Mineral fertilisers shall be applied in their natural composition and shall not be rendered more soluble by chemical treatment.

Fertilization management should minimize nutrient losses and the desired pH levels shall be maintained in the soil by the producer. Manures containing human excreta (faeces and urine) are prohibited. List of permitted, restricted and prohibited inputs are given in Annex 1 of Appendix 1 of NPOP. Products not listed in Annex 1 shall be evaluated as per the procedure given in Annex 3 of Appendix 1 and on being found complaint may be approved for use by certification body.

7. Pest, Disease and Weed Management

Organic farming systems shall be carried out in a way which ensures that losses from pests, diseases and weeds are minimized. Emphasis is placed on the use of a balanced fertilizing programme, use of crops and varieties well-adapted to the environment, fertile soils of high biological activity, adapted rotations, intercropping, green manures, *etc.* Growth and development shall take place in a natural manner. Products used for pest, disease and weed management, prepared at the farm from local plants, animals and microorganisms, shall be allowed. The use of synthetic herbicides, fungicides, growth regulators, synthetic dyes insecticides and other pesticides are prohibited. Permitted products for plant pest and disease control are listed in Annex 2 of Appendix I of NPOP. Products not listed in Annex 2 shall be evaluated as per the procedure given in Annex 3 of Appendix 1 and on being found complaint may be approved for use by certification body.

All the equipment from conventional farming systems shall be properly cleaned and free from residues before being used on organically managed areas.

8. Contamination Control

All relevant measures shall be taken to minimize contamination from outside and within the farm. Buffer zones shall be maintained to prevent contamination from conventional farms. The buffer Zone should be sufficient in size to prevent the possibility of unintended contact of prohibited substances applied to adjacent conventional land areas/farms. Polyethylene and Polypropylene or other polycarbonate coverings such as plastic mulches, fleeces, insect net and silage wrapping are allowed. These shall be removed from the soil after use and shall not be burnt on the farmland. Use of polychloride based products is prohibited.

9. Soil and Water Conservation

Soil and water resources shall be handled in a sustainable manner. Relevant measures shall be taken to prevent erosion, salination of soil, excessive and improper use of water and the pollution of ground and surface water. Clearing of land through the means of burning organic matter *e.g.* slash-and-burn, straw burning shall be restricted to the minimum. The clearing of primary forest is prohibited.

10. Collection of Non Cultivated Material of Plant Origin/Forest Produces

The collection of wild plants and parts thereof, grown naturally and in forest shall be certified as organic provided the collection areas have not received any treatment with products other than those authorised for use in organic production. The act of collection should positively contribute to the maintenance of natural areas. In harvesting or gathering the products, attention shall be paid to maintenance and sustainability of the ecosystem. Organic operators should collect products only from within the boundaries of the clearly defined wild collection area and the collection area shall be at appropriate distance from the conventional farming, pollution and contamination.

Appendix 2: Organic Livestock Standards

1. Scope

Livestock standards prescribed under these rules refer to any domestic and domesticated animal including bovine (including buffalo, Mithun and Yak), ovine, porcine, caprine, rabbits, poultry, insects and bees and/or any other animal notified by the FSSAI from time to time, raised for food/fibre or in the production of food and fibre, their derivatives and by-products. The products of hunting or fishing or wild animals shall not be considered part of livestock standards.

2. General Principles

Organic livestock production in general is a land based activity and shall be an integral part of organic farm unit and management of livestock shall be in consistent with the principles of organic farming and shall be based on:

a. Natural breeding;
b. Protection of animal health and welfare;
c. Fed with organic feed and fodder;
d. Access to grazing in organic fields;
e. Freedom to express natural behaviour;
f. Reduction of stress and
g. Prohibition of use of chemically synthesized allopathic veterinary drugs, antibiotics, hormones, growth boosters, feed additives *etc*

Landless livestock production where the operator does not have organically managed land and/or has not established a written cooperation agreement with another certified organic operator in compliance of the rules specified in Appendix 1 of these rules is prohibited.

In cases where traditional rearing system of the farm and/or adverse climatic conditions does not allow easy access to pastures, livestock may be produced through providing organic feed certified under these rules, provided the indoor and outdoor space requirements, specified under these rules are fully met (Clause 6).

3. Organic Management Plan

Any organic certification system starts with the registration of the operator with accredited certification body followed by submission of annual "Organic System plan" which provides details on all aspects of management system. Organic system plans are updated annually and all activities are carried forward throughout the year based on such organic system plan.

4. Choice of Breeds, Sources/Origin

Breeds adapted to the local conditions and resistant to diseases shall bc used. Livestock those are used for Organic production must be brought from a source under continuous Organic management.

In case if livestock has been sourced from non-organic source then they must qualify a minimum conversion period.

Certification body may allow induction of non-organic animals under certain conditions such as:

a. Establishment of new livestock farm
b. Replacement of livestock breed/strain
c. Renewal of herd necessitated by catastrophic circumstances
d. Introducing breeding males

5. Animal Identification and Record Keeping

a. Each animal/herd/batch shall bear unique identification number. Large animals including bovine, ovine, carpine, porcine *etc* shall bear individual number in the form of tag, while poultry birds and small mammals shall be identified with herd/flock/batch;
b. Following data for each animal/herd or batch shall be maintained and made available to the accredited certification body for verification during inspection:
   i. Parent details;
   ii. Source and purchase details;
   iii. Animal details;

iv. Breeding details;
v. Feeding details;
vi. Health care details including details of vaccination, medication, veterinarian prescription and withdrawal period *etc*;
vii. Production details;
viii. Sale details and
ix. Any other relevant details

6. Housing and Management

Livestock and poultry shall be maintained under natural conditions, as far as possible and shall have access to open air, grazing, runs, free range, organic feed and fodder and water.

The housing and management of the animal including sanitation, hygiene and environment shall be planned to suit the specific behavioural needs of the livestock and poultry and shall provide for:

a. Sufficient space to ensure free movement and opportunity to express normal behavior;
b. Minimum indoor and outdoor surface area requirement are given in Annex 1 of Appendix 2.
c. The animals should not be tied unless required for specific reasons
d. Appropriate facilities to cover emergencies such as the fire, the breakdown of essential mechanical services and the disruption of supplies.
e. Housing for Livestock and Poultry shall not be mandatory in areas where appropriate climatic conditions exist to enable animals to live outdoors without compromising their comfort, health and welfare.
f. Housing conditions shall meet the biological and behavioral needs of the livestock and poultry by providing easy access to feeding and watering;
g. Insulation, heating, cooling and ventilation of the building to ensure that air circulation, dust level, temperature, relative air humidity and gas concentrations are kept within limits which are not harmful to the livestock and poultry;
h. Plentiful natural ventilation and light to enter;
i. Appropriate fencing not harmful to the animals
j. Confinement allowed only under inclement weather, to ensure health safety or welfare or to protect plant, soil and water quality;
k. The stocking density shall provide comfort and well-being of the livestock and poultry with regard to the species, the breed and the age; behavioral needs with respect to the size of the group and the sex of the livestock and poultry; sufficient space to stand naturally, lie down.

7. Conversion Period

i. Simultaneous conversion of livestock and poultry and land should be a preferred approach. Land to be certified organic as per the provisions of crop production;

ii. When a livestock production unit, with entire herd, or flock is in transition to organic, pasture and feed produced on the land undergone a minimum period of 12 months of conversion period may be considered organic for feeding to organic livestock;

iii. In case of silkworm rearing, no requirement for conversion period, provided the larva are fed with organic feed grown in compliance of these rules for a minimum period of 12 months for their entire lifespan period;

iv. The conversion period shall be accounted from the date of first inspection;

v. In cases, where the land and livestock and poultry conversion to organic status is not simultaneous following conversion period shall apply

   a. Bovine including buffalo

      i. Meat products: 12 months and at least 3/4th of their life span;

      ii. Calves for meat production: 6 months;

      iii. Milk products: Six (6) months.

   b. Sheep, Goat and pigs; 6 months

   c. Small mammals (such as Rabbits) i. Meat products: From the second week after their birth to the entire life span as determined by the accredited Certification Body.

   d. Poultry

      i. Meat products: from the second day of hatching to the entire life span;

      ii. Eggs: Six (6) weeks.

8. Feed

Livestock and poultry diet shall be from feedstuffs produced as organic. Agricultural processed residues of organic origin, shall be permitted for feeding, provided overall feeding practices satisfy the daily energy and nutrient requirements.

The agriculture land for feed/fodder crops shall be organic.

During the operations, the products shall maintain their organic status provided that livestock and poultry are fed with at least 85 per cent for ruminants and 80 per cent for non-ruminants calculated on a dry matter basis, feed obtained from organic sources that have been produced in compliance with these guidelines. Accredited Certification Body can grant such permission, provided that it does not contain genetically engineered/modified organisms or products thereof.

Specific livestock and poultry rations shall take into account:

i. The need of young animals for natural feed, such as, feeding of maternal milk, milk from other mammal or milk replacer of organic origin;

ii. That in herbivores, substantial proportion of the dry matter and energy in the daily rations should consist of roughage, fresh or dried fodder, or silage; need for inclusion of cereals in the fattening phase of poultry and livestock and poultry must have ample, free access to water appropriate to maintain full health and productivity;

iii. Due to reasons of animal welfare, health and productivity, if supplements are to be added, it shall be permitted on advice of a qualified veterinarian and as listed in Annex 2.

8.1 Feedstuff and Nutritional Criteria

Substances shall be permitted as per Annex 2 and should significantly satisfy feeding requirements and such substances should not contain genetically engineered/modified organisms and products thereof; and are non-synthetic and are primarily of plant, mineral or animal origin. Accredited Certification Body may allow the use of feedstuffs not included in Annex II and have been recommended by the veterinarian, provided that all such substances are non-synthetic and are primarily of plant, mineral or animal origin.

8.2 Specific Criteria for Feedstuffs and Nutrition

i. The feedstuffs should not be prepared by using chemical solvents and chemical treatment and should be from organic sources. In case of shortage, well-defined analogic substances may also be used (Annex 3 of Appendix 2 of NPOP).

ii. Feedstuffs of animal origin, with the exception of milk and milk products, fish, other marine animals and products derived thereof shall not be used. The feeding of mammalian material to ruminants is not permitted with the exception of milk and milk products;

iii. Synthetic nitrogen or non-protein nitrogen compounds shall not be used.

8.3 Specific Criteria for Additives and Processing Aids

i. The supplements should be from natural sources

ii. Feed processing aiding supplements like binders, anti-caking agents, emulsifiers, stabilizers, thickeners, surfactants, coagulants if used should be from natural sources.

iii. Antioxidants: Only from natural sources shall be permitted

iv. Preservatives: Only natural acids are allowed;

v. Coloring agents (including pigments), flavors, odor masking agents and appetite stimulants: only natural sources are allowed.

vi. Probiotics, enzymes and microorganisms are allowed; but should not be from genetically modified sources.

vii. Any synthetic chemicals such as antibiotics, coccidiostat, medicine, growth promoters or any other substance supplemented for purpose to stimulate growth or production shall not be fed to the organic livestock and poultry.

viii. Silage additives, additives for enriching crop residues and processing aids may not be derived from genetically engineered/modified organisms or products thereof, and may be comprised of only: Sea salt; Coarse rock salt; Yeasts; Enzymes; Whey; Sugar; or sugar products such as molasses, jaggery, grain bran; Honey; Lactic, acetic, formic and propionic bacteria, or their natural acid product when the weather conditions or the fodder harvesting conditions could be perceived as a constraint to adequate fermentation provided that it is approved by the competent authority.

9. Health Care

The organic livestock and poultry, in general, should follow the basic principles of preventive health and productivity management wherein the focus would be on preventing diseases, detecting underlying fertility and production problems and its correction primarily on correcting management, nutrition and sanitation.

Draw a program of health management of animals and carry out testing of the herd as per the common diseases of herd/flock (Annex 4 of Appendix 2 of NPOP). The health care shall be based on the following broad principles:

a. The choice of appropriate breeds or strains of animals that can acclimatize, adapt to environment;

b. Setting up of animal husbandry practices appropriate to the requirements of each species and should focus on encouraging strong resistance to disease and prevention of infections;

c. Use of good quality organic feed, together with regular exercise and access to fodder/roughages, and/or open-air runs, so as to have positive effects on natural immunological defense of the animal;

d. Appropriate stocking density of livestock and poultry so as to avoid overcrowding and spread of infections or competition to feeding.

e. The farm should have an established system of detection of sub-clinical, sick or injured animals and if, so detected, must be treated immediately. In cases where isolation is necessary it will be so carried out in suitable housing areas. The paramount interest in case of sickness would be animal welfare and mitigating pain and suffering, and hence the producer shall not withhold medication even if the use of such medication will cause the animal to lose its organic status.

f. The use of veterinary medicinal products in organic farming shall comply with the following principles:

 - ☆ All vaccinations required by law of the land shall be permitted. In case there is no alternative permitted treatment or management practice exist, use of parasiticides, or therapeutic use of veterinary drugs are permitted under prescription and supervision of a registered veterinarian, provided that the mandatory withdrawal periods as provided under these guidelines are observed. In drugs where withdrawal period is not prescribed in these guidelines, a minimum of 48 hours of withdrawal period shall be observed;
 - ☆ For purpose of treatment and prevention of diseases and under-performances, herbal/phyto-therapeutic (excluding antibiotics), homeopathic or ayurvedic products shall be preferred to allopathic veterinary drugs or antibiotics, provided that their therapeutic effect is effective for the species of animal and the condition for which the treatment is intended;
 - ☆ In case alternative therapeutic or preventive measures are unlikely to be effective in combating illness or injury, allopathic veterinary drugs or antibiotics may be used under the responsibility and supervision of a veterinarian

g. The use of allopathic veterinary drugs or antibiotics or drugs derived from genetically modified source for preventative treatments and for enhancing productivity or fertility is prohibited.

h. Hormonal treatment may only be used for therapeutic reasons and under veterinary supervision.

i. Growth stimulants agents or substances used for the purpose of stimulating growth or production shall not be permitted.

10. Breeding Methods

Livestock and poultry breeding methods shall be in accordance with and in compliance with the principles of organic farming and shall take into account:

a. Breeds suited to local conditions

b. Reproduction through natural methods. Artificial insemination allowed.

c. Embryo transfer techniques and use of hormones not allowed

d. Breeding techniques employing genetic engineering not allowed

Mutilations not allowed. Certification bodies may grant some exceptions keeping in view of the health and welfare of the livestock and poultry. Physical castration is allowed to maintain the quality of products and traditional practices.

11. Manure and Excreta Management

The livestock farm shall have proper facilities for management of urine and excreta in a manner that:

- ☆ Minimizes soil and water degradation;
- ☆ Does not significantly contribute to contamination of water
- ☆ Optimizes recycling of nutrients; and
- ☆ Does not include burning or any practice inconsistent with organic practices.

All manure storage and handling facilities including composting facilities shall be designed, constructed and operated to prevent contamination of ground and/or surface water and shall be in accordance with the national standards established for the purpose.

Manure application rates shall be at levels that do not contribute to ground and/or surface water contamination. The accredited Certification Body shall establish maximum application rates for manure or stocking densities as per local conditions. The timing of application and application methods shall not increase the potential for run-off into ponds, rivers and streams.

12. Transport

During transport, the producer shall prevent stress, injury, hunger, thirst, malnutrition, fear, distress, physical and thermal discomfort, pain, disease during the transport:

- ☆ Minimize length of the journey and meet the animal's need during the journey;
- ☆ Animals must be fit for the intended journey
- ☆ Means of transport as well as the loading and unloading facilities must be designed, constructed, maintained and operated so as to avoid injury and suffering and ensure the safety of the animals;
- ☆ Personnel that handle animals must be trained and competent and ensures not to cause unnecessary fear, injury or suffering;
- ☆ Transport must carry out without delay to the place of destination and the welfare conditions of the animals must be regularly checked and appropriately maintained;
- ☆ Sufficient floor area, height and other spacing requirements must be provided for the animals, appropriate to their size and intended journey; and
- ☆ Water, feed and rest must be offered to the animals at suitable intervals and should be appropriate in quality and quantity to their species, size and age

- ☆ Efforts must be made to avoid or reduce all types of stress
- ☆ Use of electric stimulation or allopathic tranquilizers is not permitted

15. Slaughter of Animal

i. The slaughter of livestock and poultry shall be undertaken in a manner, which minimizes stress and suffering;

ii. Approved products for cleaning and disinfection of the buildings and installations are given at Annex 6;

iii. The slaughter, evisceration and packing of poultry should be conducted in such a manner as will result in hygienic processing, proper inspection and preservation for the production of clean and wholesome poultry and poultry products.

iv. Separate rooms should be provided for:
   - ☆ Live poultry receiving and holding
   - ☆ Washing and disinfection of coops.
   - ☆ Slaughter and bleeding
   - ☆ Feather removal
   - ☆ Evisceration, chilling and packing
   - ☆ Inedible products room

v. Water Supply: The quality of water should satisfy the requirements of potable water;

vi. Ventilation: Particular attention should be given to ventilation. Illumination should be sufficiently strong, properly situated and should not cause glare;

vii. Personnel hygiene: Personnel should wear special working clothes of washable material. Proper training shall be given regarding hygiene, frequent hand washing, disinfection *etc* and

viii. Activities such as stunning, bleeding, scalding, plucking, feet removal, evisceration and chilling, draining, grading *etc.* shall be done in accordance with the applicable rules framed for the purpose.

Appendix 3: Organic Apiculture Standards

1. Breeds/Strains

Preferably indigenous bee species, adapted to the local climatic conditions be used.

Sources/Origin – Preferably bee nurseries to be developed. Hive material such as boxes, wood, frames, wax *etc* are made of natural materials/sources and/or sourced from organic operations. Only natural products such as Propolis, wax and plant oils are used. Colony infested with any known disease will not be certified.

2. Conversion Period

The conversion period shall not apply when bees are grown in wild and in natural conditions. One-year conversion period shall apply to those bee colonies/ apiaries which are reared. During conversion, the bee colonies shall be placed in isolation and the foundation comb shall be made from organic wax

3. Hiving the Honey Bees

i. Where wall hives are in use, these shall accommodate movable standard frames depending upon honey bee species

ii. The foundation comb shall be made from organic wax.

iii. For renovation of apiaries, 10 per cent per year of the queen bees and the swarms may be replaced by the non-organic queen bees and swarms in the organic production unit, provided that the queen bees and swarms are placed in hives with combs or comb foundations coming from organic production unit.

iv. Each bee hive shall primarily consist of natural materials. Use of construction materials with potentially toxic effects is prohibited.

v. Persistent materials may not be used in bee hives where there is a possibility of permeation of the honey and where residues may be distributed in the area through dead bees.

vi. The apiaries shall be placed within a radius of 3 kms from the organic farms. These conditions shall not apply when the farms are not in flowering stage or when the hives are in the dormant condition.

vii. Natural products such as propolis, wax and plant oils can be used in the hives. The use of the chemical repellants is prohibited during the honey extraction operations.

4. Apiary Management

i. An apiary site shall be as close to a natural source of clean hygienic water and bee flora and well protected from weather vagaries and biotic factors.

ii. In case of wild collection, the collection area shall be organic or wild, and shall be varied as possible to fulfil the nutritional needs of the colony and contribute to good health.

iii. The number of honey bee colonies kept in such an apiary shall be limited to optimum in relation to forage resources within the same flight, range, so as to avoid over stocking.

iv. All brood or full-depth frames shall be wired to withstand breakage of combs during inspection, migration and extraction *etc.*

5. Feed

i. If needed (under unfavorable conditions), the bees can be fed with feed from organic sources

ii. Feeding shall only take place after the last harvest before the season when no foraging feed is available.

iii. At the end of the production season, hives shall be left with sufficient reserves of honey and pollen to survive the winter. The feeding of colonies shall be seen as an exception to overcome temporary feed shortages due to climatic conditions.

iv. The feeding of the colonies shall only be permitted where the survival of the hives is endangered due to climatic conditions and only between the last honey harvest and 15 days before the start of the next nectar or honey dew flow period.

6. Health Care

i. Veterinary medicine shall not be used in bee keeping.

ii. For pest and disease control and for hive disinfection the products mentioned in Annex 2 of Appendix 3 of NPOP are allowed.

iii. For protecting frames, hives and combs, in particular from pests, products listed in Annex 2 of Appendix 3 of NPOP are permitted.

iv. Physical treatments for disinfections of apiaries such as steam or direct flame are permitted.

v. The practice of destroying the male brood is permitted only if the colony is infested by Varroa destructor.

vi. If despite all the preventive measures, the colonies become sick or infested, they shall be treated immediately and, if necessary, the colonies can be placed in isolation apiaries.

7. Breeding and Management

i. Clipping of wings of queen bees are prohibited.

ii. For the renovation of apiaries, 10 per cent per year of the queen bees and swarms may be replaced by non-organic queen bees and swarms provided that the queen bees and swarms are placed in hives with combs or comb foundations coming from organic sources.

8. Periodic Cleaning

Beehives shall be cleaned periodically. Old combs shall be melted and comb renewal induced. Cleaning shall be dispensed with during the packed period.

9. Record Keeping

Records shall be maintained for each of the colonies during periodic inspections. In case of suspicion of incidence of any disease, immediate remedial measures shall be taken. The operator should maintain detailed and up-to-date records of:

- Breeding and/or origins of bees;
- Registration of any purchases;
- The health plan to be used in the prevention and management of diseases, injury and reproductive problems;
- All treatments and medicines administered for any purpose, including quarantine periods and identification of treated hives;
- Feed provided and source of feedstuffs;
- Stock movements within the unit and hive movements within designated forage areas as identified on maps;
- Extraction, processing and storing of all bee products

10. Transport/Migration

i If the local dearth period or unfavourable periods are prolonged beyond 6 to 8 weeks continuously, the colonies can be migrated to the nearest sources of organic forage from farm(s) or forest(s) as per the requirement of NSOP.

ii. Before migration, the colonies shall be packed so as to:

- Secure in position various hive components, frames in particular;
- Avoid shaking during transit;
- Provide adequate ventilation to the bees;
- Prevent congestion inside;
- Provide feeding or water in transit, if necessary; and
- Prevent honey bees escaping through gaps in entrance gates, and other components

iii. The migration shall be done preferably at night or in cool weather

iv Proper arrangement like cleaning the apiary site, arranging hive stands, providing clean water shall be done prior to the arrival of the colonies at the migratory site(s).

v. On arrival at the migratory site, the colonies shall be promptly arranged on the hive stands and the entrance gates opened at the earliest at appropriate hour.

vi. The first post–migration inspection shall be done within 7 days after the colonies settle down to work.

11. Product Extraction

i. Honey shall be extracted only when the combs are sealed by the honey bees. Extraction of unripe honey will lead to fermentation and spoilage.

ii. Honey shall never be extracted from brood combs.

iii. After honey is extracted, the empty combs shall be got cleared of honey bees and preserved carefully in supers in a cool, dry, rat-proof enclosures with suitable preservatives against wax moth and other inspect pests. Such drawn out combs shall be reused during the next honey season.

12. Extraction of Honey

i. Honey shall be extracted only from sealed combs.

ii. The use of brood combs is prohibited for honey extraction.

iii. At time of harvest, repellent consisting of prohibited substances (chemical synthetic repellents) shall not be used except smoke.

iv. Excessive smoke shall not be used.

v. Extraction shall be done only in a clean, fly-proof enclosure.

vi. All equipments used for extraction shall be cleaned in boiling water, before use. vii. During extraction, the honey shall run through a strainer of 1.40mm.

viii. The containers used for collecting the extracted honey shall be of stainless steel, aluminum or if of other metal, it shall be thickly tinned or galvanized.

ix. The container shall have covers and each shall carry a label specifying the name of the producer, date and place of extraction.

x. Persons engaged in extraction of honey shall be free from any contagious disease, shall wear clean clothes and shall clean their hands with a disinfectant soap.

xi. Honey extracted from the colonies with infectious bee diseases shall be kept separate and not mixed with general lot. This honey shall be pasteurized before marketing. It shall never be fed either in processed or unprocessed form to the bees.

xii. The extracted honey in air-tight containers shall be taken to the pooling and processing centres as early as possible. Even during the short interval the honey remains with the producer, it shall be stored in cool, dry and hygienic place and shall be protected from smoke, heat and insects

13. Extraction of Beeswax

i. Collect every bit of beeswax. Wax from different honey bee species shall be kept separately.

ii. Beeswax from cappings is the purest form of wax and shall be stored separately without mixing it with general recovery of beeswax in apiary.

iii. The old and discarded combs shall be stored in containers with tight-lids and shall be melted at the earliest to avoid further deterioration and infestation with wax moth. These melting can be cast in slabs of desired size, shape and mass.

14. Crop Pollination

A producer shall realize that besides harvesting honey and wax, he shall also mobilize his honeybees for pollination of agricultural/horticultural crops to increase the agricultural productivity.

15. Conservation of Bee Flora

Viability of the bee-keeping industry depends on the density and composition of local flora. Forest vegetation shall not, therefore, be destroyed. Trees, shrubs and herbs providing bee forage shall be particularly conserved.

Appendix 4: Organic Aquaculture Production

1. Organic Management Plan

Any organic certification system starts with the development of annual "Organic System Plan" which provides details on all aspects of management system in a prescribed format.

2. Conversion Period

The length of conversion period would vary depending upon the species, method of production, location and local conditions. Duration of the conversion period vary depending upon the species, method of production, location and local conditions and varies from minimum of six months to maximum of 24 months based upon following criteria:

a. For drainable systems where cleaning and disinfection is carried out, the conversion period shall be 6 months/one crop whichever is longer

b. In case of drainable and fallowed, the conversion period shall be 12 months.

c. In case of non-drainable systems which cannot be disinfected, the conversion period shall be 24 months (freshwater prawn, carps).

d. In case of open water farming, the conversion period shall be considered as 3 months (bivalves).

3. Ecosystem Management

Site selection and ecosystem management is done in such a way that it does not create environmentally adverse situations. Conversion of mangroves to production site and creating facilities in or through the mangroves is prohibited. The production facilities must not create the water logging in the adjacent areas and must not contribute to salination, nutrient discharge or pollution of water bodies and must not undermine the diversity.

A buffer zone of at least 10 m should be left between farms following organic farming principles and conventional farming. The size of the buffer zone could be increased based on the natural situation, water distribution system, tidal flow and the upstream and downstream locations of organic production unit.

4. Selection of site

i. In selecting the site, ensure that the surrounding aquatic and terrestrial ecosystems are not adversely affected through modifications brought about by construction of the farm.

ii. Areas with known record of contamination with heavy metals or industrial pollution may be avoided.

iii. Soil quality should be conducive for culture and extreme conditions like high saline or acid soil may be avoided.

iv. Use of ground water for the culture purpose of tiger shrimps is prohibited. For other species the groundwater should be avoided.

v. In case of the bivalve farm, the location of the farm should be as close as possible to the sea to ensure maximum circulation of sea water.

5. Choice of Breeds

Endemic species be preferred over exotic species. Use of genetically engineered stock is prohibited.

6. Source of Seed and Breeding

Breeds and the breeding techniques appropriate for the species, environment, production systems and local conditions should be followed for minimizing stress to the brood stock. As far as possible, only organic source breed stock should be used. When organic seed is not available, the certifying body may prescribe a time limit for use of non-organic seed depending upon the species. For carps and fresh water prawns, the maximum percentage of non-organically produced juveniles allowed to be introduced to the farm shall be 80 per cent, 50 per cent and 0 per cent by second, third and fifth year from the year of notification.

Uses of synthetic chemicals/hormones/antibiotics are not allowed in breeding. Since exogenous hormone supply is an essential requisite for induced spawning of carps, use of pituitary gland may be accepted.

In case of bivalves, collection of natural brood stock is permitted but use of chemicals as a means of triggering spawning is not allowed.

7. Culture Practices

Husbandry practices, including feeding, design of installations, stocking densities and water quality shall ensure that the developmental, physiological and behavioural needs of animals are fully met

8. Pond Preparation

Ponds are disinfected materials and fertilization to be done by organic sources comprising mainly of biodegradable nature from plant/animal origin with physical or organically acceptable.

Integrated farming system can be adopted for recycling of nutrients. Cow dung/poultry manure/farm yard manure/vermi-compost may be used as nutrient source for carp farming. Intermittent application of cow dung/poultry manure during culture operation should be in the fermented form. The manure to be used should be from organic sources.

9. Stocking

Stocking density is maintained in such a way so as not to compromise with the animal well being, ecological capacity of the site and species specific physiological need and animal behaviour. Limits of maximum stocking density are as follows:

a. For shrimp farming 6 nos/sq mt and biomass in the pond shall not exceed 1400 kg/ha/crop
b. For freshwater prawns up to 2.5 nos/sq mt and biomass in the pond shall not exceed 800 kg/ha/crop.
c. For nursery rearing of freshwater prawns 20 nos/sq mt is permitted.
d. For carp fry and fingerlings production in nursery 2 million spawn/ha (200 nos/sq mt) and 0.1 million fry/ha (10 nos/sq mt), respectively.
e. For grow-out production of carps 4000 fingerlings/ha (0.4 nos/sq mt) may be followed and the maximum biomass should not exceed 3 tonnes/ha at any point of time.
f. In case of carp farming, polyculture of compatible carp species is preferred over monoculture in order to utilize the ecological niche effectively.

10. Pond Management

Ponds are required to be designed to maintain suitable environment, most befitting with the natural behaviour of the stock. The water quality must be conducive for the species to live in. The energy requirement for mechanical operations be kept minimum.

For cleaning and disinfections, only substances from approved list shall be used. Measures of aeration must not be used in the pond to raise the stocking density above the permitted level.

As far as possible avoid use of plastics except for most essential items such as nets, crates, floats *etc.*

11. Bivalve Farming

In the case of bivalves like mussels and oysters, the grow-out methods permitted are off bottom racks, rafts, long-lines and stakes using ropes and nets. Production shall take place within areas delimited by posts, floats or other clear demarcations and shall as appropriate be restrained by net bags, cages or other man made means. In case of mussels, the stocking density should not exceed 2 kg/m rope and the production should not exceed 15 kg/m rope.

12. Supplementary Feeding

Natural productivity of the pond should be the main source of feeding and external feeding to be resorted as supplementary in the form of farm made feeds from organic sources. All feed ingredients must be from organic and mineral sources. Ingredients from GMO sources are prohibited.

In case of tiger shrimp and freshwater prawn, the fish meal content in the feed should not exceed 20 per cent and the total protein content of animal origin should not exceed 25 per cent.

In case of carp farming, use of animal protein including fish meal in supplementary feed should be avoided. Feed prepared from certified organic ingredients is to be used for supplementary feeding. Use of antibiotics/pesticides/heavy metals/antioxidants/preservatives/growth hormones *etc* is prohibited.

Minerals, trace elements, vitamins or pro-vitamins to be used in the feed shall be of natural origin. Growth promoters and synthetic amino acids are not permitted.

Culture of live fish food organisms like algae, rotifers, artemia *etc* for shrimp hatchery may be carried out in accordance with principles of organic agriculture.

13. Health Management

Prevention is better than cure should be the principal for health management. Use of human excrement, sewage, sludge and chemotherapeutics with allopathic veterinary drugs are prohibited. Yeast based preparations and probiotics of certified organic origin are allowed. GMO based preparations are not allowed.

14. Harvest and Transport

Harvesting method shall be human and aquatic animals shall be subject to minimum stress during harvest. Animals sold live should be transported with minimum stress. Others should be chill killed at farm site itself. Use of chemicals like sodium metabisulphate is prohibited; however ascorbic acid is allowed to stop discoloration (Refer Annex 2 and 3 of Appendix 4 of NPOP).

15. Processing

Processing and packaging of the organic produce shall be carried out in the Organic certified processing units. Defined measures shall be taken to maintain the organic integrity of the processed product. The limit of permitted and prohibited substances for use in aquaculture processing shall be as per Annex 5 of Appendix 4 of NPOP.

Appendix 5: Organic Food Processing and Handling

1. Organic Food Processing and Handling

Plan Any handling and processing of organic products should be optimized to maintain the quality and integrity of the product. The operator must develop an organic production and handling plan. This plan shall include:

i. Description of practices and procedures to be performed
ii. List of each substances/inputs used during production, storage and handling. Description of the monitoring practices and procedures followed
iii. Description of the monitoring practices and procedures followed and maintained.
iv. Description of record keeping system.
v. Description of management practices and separation measures established to prevent mixing and co-mingling.
vi. Pollution sources shall be identified and contamination avoided.
vii. Processing and handling of organic products should be done separately.
viii. All products shall be adequately identified through the whole process.
ix. Certification programme shall regulate the means and measures to be allowed.

x. Recommended for decontamination, cleaning or disinfections of all facilities.

2. Pest Control

i. Pest should be avoided by good manufacturing practices such as general cleanliness and hygiene.
ii. Treatments with pest regulating agents must thus be regarded as the last resort
iii. Recommended treatments are physical barriers, sound, ultra-sound, light and UV-light, traps (incl. pheromone traps and static bait traps), temperature control, controlled atmosphere and diatomaceous earth
iv. A plan for pest prevention and pest control should be developed
v. For pest management and control the following measures shall be used in order of priority:
   - Preventive methods such as disruption, elimination of habitat and access to facilities
   - Mechanical, physical and biological methods
   - Pesticidal substances contained in the Appendices of the national standards
   - Other substances used in traps
vi. Irradiation is prohibited.
vii. There shall never be direct or indirect contact between organic and prohibited substances.
viii. Persistent or carcinogenic pesticides and disinfectants are not permitted.

3. Ingredients

100 per cent of the ingredients used in processing shall be organic except where an organic ingredient is not available in sufficient quality or quantity and non organic ingredients may be used to a minimum extent only in case of essential technological need or for particular nutritional purpose which shall not be genetically engineered. The accredited Certification Body may authorize the use of non-organic raw materials subject to periodic re-evaluation.

Water and salt may be used and Minerals (including trace elements), vitamins and similar isolated ingredients shall not be used. The certification programme may grant exceptions where use is legally required or where severe dietary or nutritional deficiency can be demonstrated. Ethylene gas is permitted for ripening.

4. Processing Methods

The vital quality of an organic ingredient shall be maintained throughout each step of its processing methods and shall be chosen to limit the number and quantity of additives and processing aids. The approved kinds of processes include mechanical and physical, biological, smoking, extraction, precipitation and filtration.

Extraction shall be either with water, ethanol, plant and animal oils, vinegar, carbon dioxide, nitrogen or carboxylic acids. Filtration substances shall not be made of asbestos and Irradiation is not allowed.

5. Packaging

Biodegradable, recyclable, reusable systems and eco-friendly packaging materials shall be used wherever possible. Certain additives for use in manufacturing of packaging films for packaging of organic food stuffs are allowed for restricted use are listed in Annex II of Appendix 5 of NPOP. The accredited Certification Body shall approve the packaging material for use.

6. Labeling

i. Labeling shall convey clear and accurate information on the organic status of the product.

ii. When full standards requirements are fulfilled, products shall be sold as "produce of organic agriculture" or a similar description.

iii. The label for conversion products shall be clearly distinguishable from the label for organic products by mentioning the year of conversion.

iv. The name and address of the person or company legally responsible for the production or processing of the product shall be mentioned on the label.

v. Product labels should list processing procedures which influence the product properties in a way not immediately obvious. All components of additives and processing aids shall be declared.

vi. Additional product information shall be made available on request.

vii. Ingredients or products derived from wild production shall be declared as such.

Processed Products

i. Single ingredient products may be labelled as "Organic" when all standard requirements have been met.

ii. Multi ingredient products where not all ingredients including additives are of organic origin may be labelled in the following way (raw material weight):

- ✰ Where a minimum of 95 per cent of the ingredients are of certified organic origin, products may be labelled "certified organic" or similar and should carry the logo of the certification programme.
- ✰ Where less than 95 per cent but not less than 70 per cent of the ingredients are of certified organic origin, products may not be called

"organic". The word "organic" may be used on the principal display in statements like "made with organic ingredients" provided there is a clear statement of the proportion of the organic ingredients. An indication that the product is covered by the certification programme should be used, close to the indication of proportion of organic ingredients.

- ✰ Where less than 70 per cent of the ingredients are of certified organic origin, the indication that an ingredient is organic may appear in the ingredients list. Such product may not be called "organic".

iii. Added water and salt shall not be included in the percentage calculations of organic ingredients. For aquaculture products, the use of iodized salt shall be referred on the labels.

iv. All raw materials of a multi-ingredient product shall be listed on the product label in order of their weight percentage. It shall be apparent which raw materials are of organic certified origin and which are not. All additives shall be listed with their full name.

v. f herbs and/or spices constitute less than 2 per cent of the total weight of the product, they may be listed as "spices" or "herbs" without stating the percentage.

vi. Organic products shall not be labeled as GE (genetic engineering) or GM (genetic modification) free in order to avoid potentially misleading claims about the end product. Any reference to genetic engineering on product labels shall be limited to the production method.

vii. The label of a certified organic product must depict the name and logo of the accredited Certification Body, accreditation number and India Organic Logo viii. The accredited Certification Body shall verify the labeling requirement and approve the labels of their certified operators before the labels are used.

7. Storage and Transport

i. Organic products shall be stored at ambient temperature. The special conditions of storage permitted are: Controlled atmosphere, Cooling, Freezing, Drying, Humidity regulation.

ii. Product integrity should be maintained during storage and transportation of organic products. Organic Products must be protected at all times from commingling with non-organic products and from contact with materials and substances not permitted for use in organic farming and handling.

iii. In units processing organic and non-organic, the organic products should be handled and stored separately.

iv. In bulk stores organic products should be separate from conventional

v. Storage areas and transport containers for organic product should be cleaned using methods and materials permitted in organic production. Measures should be taken to prevent possible contamination from any pesticide or other treatment not listed in Annex-2 of Appendix I of NPOP.

Appendix 6:
Organic Animal Feed Processing and Handling Standards

Organic animal feed processing and handling includes processing of organic feed and food for all types of domesticated animals including livestock, poultry, aquaculture and pet animals for production of commercial animal feed/food products.

1. General Requirements

(i) The handling and processing operator of organic animal feed/food products must set up a written organic handling plan.

(ii) Necessary measures shall be put in place to minimize air, water and soil contamination.

(iii) Description of the monitoring practices and procedures

(iv) Description of the record keeping system

(v) Description of the management practices and separation measures

2. Cleaning, Disinfection and Pest Control

(i) Preventive measures need to be put in place to protect organic feed from substances prohibited for use in production, processing, manufacture or handling from pests, pathogens and other alien substances.

(ii) The cleaning and disinfection of facilities and sterilization of equipments and tools substances are listed in Annex 9 of Appendix 2 of NPOP and Organic feed must not come in their contact.

(iii) The measures for pest management and control are mentioned hereunder:

- ☆ Preventive methods such as disruption, elimination of habitat and access to facilities
- ☆ Mechanical, physical and biological methods
- ☆ Physical barriers, sound, ultra-sound, light and UV-light, traps (incl. pheromone traps and static bait traps), temperature control and controlled atmosphere
- ☆ Pesticidal substances contained in the Appendices of the national standards.
- ☆ Irradiation is prohibited and Persistent or carcinogenic pesticides and disinfectants are not permitted

3. Ingredients

(i) The ingredients and supplementary feed used for production of organic feed shall be derived from following sources:

- ☆ Organic crop products
- ☆ Organic livestock product
- ☆ Organic processed products
- ☆ Ingredient or supplementary feed referred to in Annex 5 and Annex 6 of Appendix 2 and in Annex 2 and 5 of Appendix 4 of NPOP.

(ii) In case of non-availability of an organic ingredient, non- organic ingredients may be used to a minimum extent only in case of essential technological need. Accredited certification bodies may allow the use of nature identical synthetic amino acids and vitamins in cases where their requirement cannot be met by other permitted sources.

(iii) Preparations of micro-organisms and enzymes commonly used in food processing may be used with the exception of genetically engineered micro-organisms and their products. Water and salt may be used in organic feed processing.

(iv) Use of Genetically modified organisms, Synthetic chemicals used for boosting metabolism, Synthetic nitrogen or non-protein nitrogen compounds, Antibiotics, Synthetic antimicrobials, Growth enhancing substances, Parasiticides, Hormones, Feed or raw material of mammalian origin including slaughter house waste for making feed for ruminant livestock *etc.* are prohibited.

4. Processing

Processing methods should be based on mechanical, biological, smoking, extraction, precipitation and filtration. Water, ethanol, plant and animal oils, vinegar, carbon dioxide, nitrogen or carboxylic acids may be used for extraction. Filtration substances shall not be made of asbestos nor may they be permeated with substances which may negatively affect the product. Irradiation is not allowed.

5. Processing Facilities

Processing facilities to be managed in such a way that organic integrity is maintained throughout the process without any chance for mixing or co-mingling with nonorganic products or ingredients. Organic feed production lines must be separated from non-organic feed production line and Separate storage facilities must be in place and managed separately so that ingredients used for producing organic feed do not get mixed with non-organic ones.

6. Processed Products

A raw or processed animal feed sold, labelled or represented as "100 per cent organic" and "Organic" it must contain 100 per cent organically produced ingredients and not less than 95 per cent organically produced raw or processed agricultural ingredients (by weight or fluid volume, excluding water and salt), respectively. Multi-ingredient animal feed sold, labelled or represented as "made with organic (specified ingredients or food group)" must contain (by weight or fluid volume, excluding water and salt) at least 70 per cent organically produced ingredients. If less than 70 per cent of the ingredients are of certified organic origin, then the product may not be called "organic".

7. Packaging

Packaging methods and materials must protect the integrity of organic feed and have no adverse effects on the environment. Biodegradable, recyclable, reusable systems and eco-friendly packaging materials shall be used. Packaging materials, containers and storage containing or treated with synthetic chemicals or prohibited substances must not be used. Recycled packaging materials or containers that had come in contact with substances that may compromise the organic integrity of organic feed must not he used and the packages shall be closed in such a manner that substitution of the content cannot be achieved without manipulation or damaging the seal.

8. Labeling

All organic processed animal feed shall be labelled as per the requirements specified in Clause 6 (6.1 and 6.2) of Appendix 5 of NPOP.

9. Storage and Transport

Storage, shipping and transportation requirements must be in compliance of Clause 7 of Appendix 5 of NPOP.

Appendix 7: Organic Mushroom Production

1. General

Mushroom production is although similar to crop production but with a difference that it is an indoor activity under controlled environment and without the involvement of soil as growing medium. Mushroom production standards cover all edible mushrooms intended for human consumption, whether grown on compost, raw biomass or wood.

2. Organic Management Plan

During the registration of the farm or organic mushroom production unit with the Accredited Certification Body (ACB), the operator has to submit an organic management plan, which will be verified by the ACB during inspection. The organic management plan shall be updated annually.

3. Management of Production Site

The operator shall maintain the entire production site including housing facilities in a way that prevents contact with prohibited substances with production site, tools and boxes/trays, organically produced mushrooms and each and every step throughout the entire growing cycle including harvesting and post-harvesting process. Any wood or plant material used for construction of mushroom house, racks, substrate holding containers, boxes, trays *etc* shall be free from prohibited substance treatment.

Organic and non-organic production units must be in separate facilities separated by space and time and have separate ventilation systems, boxes, trays, tools, substrate holding racks *etc.* including facilities for compost production.

4. Substrate and Growing Media

All substrate and growing media shall be prepared on the farm in compliance of these standards or sourced from certified organic sources certified in accordance with the standards prescribed in Appendix 1 of these rules.

In case of unavailability of certified organic raw material needed for making the substrate accredited certification bodies may allow the use of chemically untreated conventionally grown raw material up to a maximum limit of 25 per cent for making the compost. The composting process shall ensure that the substrate has reached a temperature of at least 65^{0}C for about 6-7 days prior to use. All composts and growing media used (from the commencement of the composting process) shall be audited and verified for compliance with this Standard by the accredited certification body. Steam is allowed for final sterilisation of compost.

In cases where raw crop residue/biomass is used without composting as substrate, such as straw, hey or grains, they shall be sourced from organic operations certified as per crop production standards prescribed under Appendix 1 of these rules.

Logs, sawdust or other wood based material when used as substrate shall come from wood, trees or logs that have not been treated with prohibited substances.

5. Fungus Spawn

Organic spawn (seed) shall be used. Accredited certification bodies shall evaluate the conformance of spawn production as per the evaluation process given in Annex 3 of Appendix 1 under these rules. In case of non-availability of organic spawn accredited certification bodies may allow the use of conventionally grown spawn for limited period of time.

Use of GMO products or its derivatives or genetically modified organisms (spawn) at any stage of the production process is prohibited.

6. Conversion

Existing Mushroom production systems on being converted to organic management shall have to undergo a minimum period of 12 months as conversion period from the date of registration with the certification body. During the conversion period all management practices must be in compliance of these standards.

In case of new installations where the entire production system is being implemented in compliance of these standards, two or more production cycles must have been produced under organic conditions compliant with this standard prior to products being sold as organic.

7. Pest Control and Sanitation

Preventive pest and disease management shall be the preferred approach. Methodologies and measures listed at Clause 8 of Appendix 1 and at Clause 2 of Appendix 5 can be used in cases where preventive measures are not sufficient to tackle the problem.

For sanitation and disinfection of installation, equipments and facilities products listed in Annex 9 of Appendix 2 of these rules can be used.

Appendix 8: Organic Seaweed, Aquatic Plants and Greenhouse Crop Production

1. General

Organic seaweed, aquatic plants (including algae) and greenhouse crop production being crop production activity, needs to comply the overall requirements, unless otherwise described under these rules as exception, of crop production rules prescribed under Appendix 1 of these rules.

Organic Seaweed

Organic seaweed production includes collection of wild seaweeds and parts thereof growing naturally in the sea and cultivated in the coastal areas for use as food for human or livestock consumption or for use as raw material for processing of food or feed.

Organic Aquatic Plants

Organic aquatic plants includes macro and micro green plants including algae grown under aquatic environment in open natural habitat or under artificial conditions in ponds or tanks in open or under greenhouse conditions. Greenhouse crops Greenhouse crops includes general agricultural and horticultural crops cultivated under greenhouse conditions in permanent in-ground soil systems or in containers filled with plant and soil based growing substrate connected with soil, except nursery plants which can be grown in containers in plant based growing medium.

2. Organic Management

Plan During the registration of the farm or production site/unit with the Accredited Certification Body (ACB), the operator has to submit an organic management plan, which will be verified by the ACB during inspection. The organic management plan shall be updated annually.

3. Specific Requirements for Seaweeds

3.1 Collection from Wild

The collection of wild seaweeds and parts thereof shall comply with the overall requirements specified under Clause 11 of Appendix 1 as applicable under sea ecosystem. In addition the wild sea weed collection shall also be subject to following:

i. The collection area shall be far away from human habitation and human activity and free from any external contamination source.
ii. The collection area shall be of sound ecological quality and not declared unsuitable from human health point of view.
iii. The collection shall not affect the long term sustainability of the natural habitat or the maintenance of the species growing in the area.

3.2 Cultivation in Sea and Inland Tanks

The cultivation of sea weeds can be taken up in coastal areas under natural conditions or under inland tanks with specific purpose. Following specific rule shall be followed in sea weed cultivation:

i. Coastal area where sea weed cultivation is done must be free from any external contamination source and at a distance from human habitat.
ii. The cultivation area shall be of sound ecological quality and not declared unsuitable from human health point of view.
iii. Sustainable practices leading to natural conditions be used in all stages of production starting from collection of juvenile sea weed to harvest.
iv. Seeding of seaweeds can be done by indoor culture stocks grown under conditions specified in these rules.
v. In case of non-availability of organic seed material and/or to maintain the wide gene pool with natural vigour, juvenile sea weed from the wild can be supplemented in the growing area.
vi. No fertilizers or any growth enhancing input shall be used in natural cultivation area on the coasts.
vii. In case if seaweed is cultivated in tanks or juvenile seaweeds are raised in tanks then the coastal marine water without any treatment be used and

the tanks shall have bottom surface as natural soil. Cultivation of seaweed in complete cemented tanks or made of artificial material without any contact with soil is prohibited.

viii. Under inland tank conditions inputs authorised for use in crop production under Annex 1 and Annex 2 in Appendix 1 can be allowed by the certification body.

ix. Use of synthetic inputs such as fertilizers, pesticides, hormones *etc* and genetically modified organisms or their products are prohibited.

x. For sanitation and hygiene maintenance of tanks inputs allowed under Annex 9 of Appendix 2 can be authorised by the Certification body, but in all such cases it must be ensured that the washings of such operations are not drained to the sea.

xi. In areas where cultivation is done in sea coast the product shall be allowed to be sold as organic after a minimum period of six months after the date of first inspection by the certification body. In case of inland tanks the product shall be allowed for sale as organic only after 24 months of starting the production after the date of registration with the accredited certification body. In cases where operator can demonstrate to the satisfaction of ACB that the land where cultivation tanks have been made has not been used for any cultivation activity then the conversion period can be reduced to 12 months after the date of first inspection.

xii. Organic and non-organic production units must be in separate facilities separated by space and time and have separate equipments, storage, processing facilities and drying beds. Tanks used for cultivation of seaweeds with prohibited inputs shall not be used for cultivation of organic seaweeds unless have gone through the conversion period as mentioned above.

4. Aquatic Plants including Algae

Cultivation of aquatic plants is a crop production activity and all requirements under Appendix 1, crop production, of these rules as applicable under aquatic environment shall apply including conversion requirements.

i. Cultivation of aquatic plants in artificial tanks without any soil base or organic substrate/media complying to the standards does not qualify for organic production under these rules.

ii. Organic and non-organic production units must be in separate facilities separated by space and time and have separate equipments, storage, processing facilities and drying beds. Tanks used for cultivation of aquatic plants with prohibited inputs shall not be used for cultivation of organic aquatic plants unless have gone through the conversion period as mentioned above.

iii. The water used for cultivation shall be of potable quality and the soil shall be free from any contamination including heavy metals.

iv. Use of synthetic chemicals/prohibited substances for sterilization/ sanitation of production sites is prohibited, except the ones allowed under these rules.

v. Mother culture or seeding material shall also be organic in compliance of these rules. In case of non-availability, non-organic seeding material can also be used without any chemical treatment or contamination.

vi. Use of genetically modified seeding material is prohibited

vii. Weeds shall be controlled by physical or biological prevention methods

viii. Use of chemical fertilizers (including trace elements), pesticides, hormones *etc.* is prohibited

ix. Mineral fertilizers in their natural composition can be used. Fertilization practices shall be in conformity of practices allowed under Appendix 1 of these rules.

x. Physical and biological practices can be used for pest management. Use of synthetic chemical substances and plants extracts harmful to human health shall not be used.

xi. Inputs or substances approved under these rules in Appendix 1 can be used with the prior permission of certification body.

xii. Processing of aquatic plants and their parts thereof shall be done in accordance with the requirements specified under Appendix 5 of these rules.

5. Greenhouse Crop Production

Greenhouse crop production is a crop production activity with difference that it is done under partially controlled conditions. All the requirements specified under Appendix 1 shall also apply under greenhouse (Glass house, poly house or net house) conditions, including conversion requirements of land. In addition following requirement shall also be met:

i. Greenhouse design and its surroundings shall be orientated towards environmentally positive outcomes and resource efficiency, including water reuse where applicable.

ii. Hydroponic and aeroponic systems where plants are fed principally through soluble fertilisers through water cannot be certified under these rules as they are not grown in healthy and complex soil ecology.

iii. Under greenhouse conditions parallel production or split production under same greenhouse is prohibited. In case if an operator cultivate both organic and conventional crops under greenhouse then the two systems must be separate with adequate buffer zone and implements/

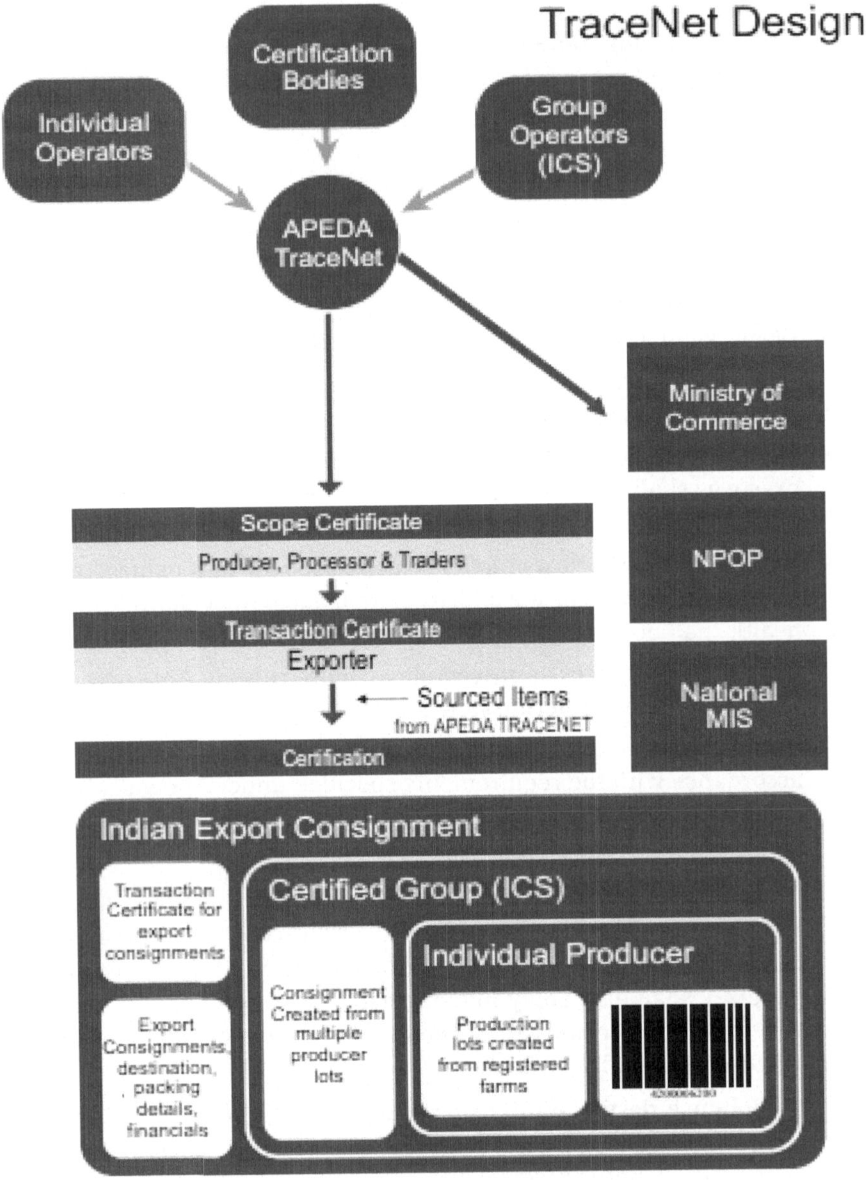

equipments used must be properly cleaned before using under organic operations.

iv. During non-crop-production periods, a cover crop or green manure phase, or similar methods, shall be practised to ensure ongoing soil life protection and enhancement.

v. Media used to produce plants may include coconut fibre and other sources permitted under Annex 1 of Appendix 1 of these rule or have been evaluated for their suitability as per the procedure prescribed under Annex 3 of Appendix 1 of these rules.

vi. Media shall have contact with soil or mixed with soil and shall be incorporated or recycled during or at the end of the cropping cycle.

vii. Where containers are used, containers shall consist of non-contaminating products of plant origin. Optimally such containers shall be reusable after phytosanitary considerations are satisfied.

viii. Sterilisation of growing containers for purposes of disease management shall either utilise steam, heat or other physical means or other practices or products listed in Annex 9 of Appendix 2 of these rules.

ix. The fertility management shall be in accordance with the fertilization policy for crop production under these rules.

x. A diversity of crop species shall be chosen in any one season to ensure good rotations and general diversity.

xi. Intercrops and harbouring floral species shall be encouraged for biocontrol agents.

xii. Heating and lighting, where used, shall achieve best management practice in terms of efficiency and environmental impact, and wherever practicable shall rely upon renewable resources.

Glossary

Abiotic: Non-living. Abiotic resources comprise nonliving things, for instance, land, water, air and minerals.

Accreditation: Accreditation means Registration by the National Accreditation Body for certifying organic farms, products and processes as per the National Standards for Organic Products and as per the guidelines of the National Accreditation Policy and Programme for organic products.

Accredited Programme: Means programme of accrediting Inspection and Certification Agencies which have been accredited by the Accreditation Agency and which have agreed to comply with the Accreditation contract.

Adapted to local conditions: The ability of organic farming, pastoral and wild harvest systems to fit into the cycles and ecological balances in nature. Organic management must be adapted to local conditions, ecology, culture and scale of operations.

Adaptive management: Management practices that promote a system's ability to take advantage of opportunities or cope with problems occurring in the environment. Considering the high level of variability within and between ecosystems, and the reliance of organic agriculture on local ecological balance, adaptive management is a central strategy. In organic agriculture, uncertainty does not only apply to biophysical variability but also to lack of knowledge and advisory services necessary to improve agricultural performance. Spontaneously, and in "learning by doing", organic farmers improve their management approach and through observation and experimentation, they determine the best management strategy within their own context, including available capabilities, resources and institutions.

Agricultural biodiversity: The component of biodiversity that is relevant to food and agriculture production. The term agrobiodiversity encompasses genetic species and ecosystem diversity.

Agricultural biomass: Non-fossil biological material, either from plant or animal origin, both living and dead, found above or below ground vegetation, including agricultural products and waste by-products, manure, soil fauna or microbial biomass used as food, feed, fuel or for soil amendment.

Agricultural by-product: It includes, among others, maize cobs and stalks, wheat stalks and husks, groundnut husks, cotton stalks, mustard stalks, *etc*. Definition: Vegetal or animal material and by-product derived from production, harvesting, transportation and processing in farming areas.

Agricultural ecosystem: A semi-natural or modified natural system managed by humans for food and agricultural production purposes.

Agricultural intensification: Refers to any practice that increases productivity per unit land area at some cost in labour or capital inputs. One important dimension of agricultural intensification is the length of the fallow period (i.e. letting land lie uncultivated for a period) and whether the management approach uses ecological or technological means.

Agricultural product: Any product or commodity, raw or processed, that is used for human consumption (excluding water, salt and additives) animal feed or energy production (*e.g.* biofuel crops) and industry (*e.g.* textiles, bioplastics).

Agriculture value-added: Annual growth rate for agricultural value-added based on constant local currency. Aggregates are based on constant 1995 U.S. dollars. Agriculture corresponds to the International Standard Industrial Classification (ISIC) divisions 1-5 and includes forestry, hunting, and fishing, as well as cultivation of crops and livestock production. Value added is the net output of a sector after adding up all outputs and subtracting intermediate inputs. It is calculated without making deductions for depreciation of fabricated assets or depletion and degradation of natural resources.

Agri-environmental measure: Special measures applied in order to promote the protection of the farmed environment and its biodiversity. Agri-environmental measures support specifically designed farming practices, going beyond the baseline level of "good farming practice" that help to protect the environment and maintain the countryside.

Agritourism: Agritourism is a style of vacation in which hospitality is offered on farms. Agri-tourism can refer to various kinds of small farms seeking to diversify their enterprises to strengthen their financial position.

Agrochemical: Agrochemicals are commercially produced, usually synthetic, chemical compounds used in farming such as a fertilizer, pesticide or soil conditioner. In organic agriculture, agrochemicals are banned and any use of substances for soil fertilization and conditioning, pest and disease control, for the health of livestock and quality of the animal products, or for preparation, preservation and storage of the food product should comply with the relevant national regulations.

Agro-ecological knowledge: Ecological knowledge refers to what people know about their natural environment, based primarily on their own experience and observation. Agro-ecological knowledge refers to farmers' knowledge of ecological interactions within the farming system.

Agro-ecology agroecology: Agroecology is the science and practice of applying ecological concepts and principles to the study, design and management of the ecological interactions within agricultural systems (*e.g.* relations between and among biotic and abiotic elements). This whole-systems approach to agriculture and food systems development is based on a wide variety of technologies, practices and innovations including local and traditional knowledge as well as modern science.

Agro-ecosystem stability: In general, stability (of the ecosystem) refers to the capability of a natural system to apply self-regulating mechanisms so as to maintain its balance in the face of an outside disturbance.

Agro-ecotourism: Eco-agritourism combines rural tourism (agritourism) and ecological tourism (eco-tourism) with farm hospitality and enjoying neighbouring natural landscapes.

Agroforestry: A collective name for land-use systems and technologies where woody perennials (trees, shrubs, palms, bamboos, *etc.*) are deliberately used on the same land management unit as agricultural crops and/or animals, in some form of spatial arrangement or temporal sequence. In agroforestry systems, there are both ecological and economic interactions between the different components.

Animal immunity: The state of relative insusceptibility of an animal to infection by disease-producing organisms or to the harmful effects of their poisons.

Annual Report: Means the report on operators, products and processors submitted annually to the Accreditation Agency by the accredited Inspection and Certification Agencies.

Antibiotic-free: When no antibiotic drugs have been given to the animal in its feed or by injection.

APPEAL: Shall be the process by which an Inspection and Certification Agency can request reconsideration of a decision taken by the Accreditation Agency or an operator can request reconsideration of a decision by the Certification Agency.

Applicant: Shall be the Inspection and Certification Agency that has applied for Accreditation to the Accreditation Agency.

Autochtonous: Existing, born, or produced in a land or region.

Avoidance costs: The actual or imputed costs of preventing environmental deterioration by alternative production and consumption processes, or by reduction of or abstention from activities.

AYURVEDA: Ayurveda is a traditional naturopath system of medicines and health care of India.

Beneficial association: An association of plants, animals and microorganisms is called beneficial when complementarity is achieved with respect to nutrient and energy up-take. Associations can take different forms: from companion planting to arrangements between members of different species such as symbiosis and parasitism.

Biodiversity: The variability among living organisms from all sources including, inter alia, terrestrial, marine and other aquatic ecosystems and the ecological complexes of which they are part; this includes diversity within species, between species and of ecosystems.

Biodynamic agriculture: Biodynamic agriculture considers both the material and spiritual context of food production and works with terrestrial as well as cosmic influences. The influence of planetary rhythms on the growth of plants and animals, in terms of the ripening power of light and warmth, is managed by guiding cultivation times with an astronomical calendar. All organic principles apply to biodynamic farming, gardening and forestry.

Bio-ecotourism: Refers to ecological tourism combined with organic agritourism, where organic farm services include organic food and hospitality in buildings constructed with ecological materials and farm infrastructures including environmentally-friendly structures such as waste recycling, renewable energy and other environmentally friendly structures.

Bio-educational farm: An organic farm dedicated to educational activities. A bio-educational farm is committed not only in ensuring naturalistic services to guests but also didactical ones in order to make people participate and have a deeper knowledge of organic lifestyles, including tending animals, cooking and artisanal crafting.

Biofertiliser: A biofertilizer is a natural fertilizer that helps to provide all the nutrients required by the plants and to increase the quality of the soil with a natural microorganism environment. For instance, the production and use of biofertilizer (*e.g.* seaweed products; compost) is proposed to improve crop yields by using root nodule bacteria (rhizobia), mycorrhizal fungi, and other microorganisms that are able to increase the accessibility of plant nutrients from the soils. Related Terms: nitrogen fixation; biocontrol agent; biological pest control; organic fertilization; farmyard manure; organic manure; compost; humus.

Biological pest control: Biological control is a method of controlling pests, diseases and weeds in agriculture that relies on natural predation, parasitism or other natural mechanisms that restrain the development of pathogenic organisms. The control of living organisms (especially pests) by biological means. Any process using deliberately introduced living organisms to restrain the growth and development of other, very often pathogenic, organisms, such as the use of spider mites to control cassava mealybug. The term also applies to the use of disease-resistant crop cultivars. Biotechnology approaches biocontrol in various ways, such as using fungi, viruses or bacteria, which are known to attack an insect or weed pest.

Biologically-based product: Biologically-based products are liquids, powders and/or granules that contain as their main active ingredient any source of beneficial microbes (bacteria, fungi, *etc.*) that help protect the plant from pests and diseases and/or help to enhance plant growth. These types of products include: biopesticides, inoculants, soil conditioners, biostimulants, *etc*

Biomass: The total weight of all the biological material or the combined mass of all the animals and plants inhabiting a defined area; usually expressed as dry weight per area.

Buffer zone: A clearly defined and identifiable boundary area bordering an organic production site that is established to limit the application of, or contact with, prohibited substances from an adjacent area.

Carbon capture: Conversion, through photosynthesis, of atmospheric carbon leading to the long-term storage of carbon in the soil and in living and dead vegetation. Carbon stored can offset carbon dioxide released. Therein lies the possibility of agriculture providing a valuable service to society by storing carbon that offsets the carbon dioxide that is emitted by other sectors.

Carbon trading: is a form of emissions trading that allows a country to meet its carbon dioxide emissions reduction commitments, often to meet Kyoto Treaty requirements, in as low a cost as possible by utilising the free market. It is a means of privatising the public cost or societal cost of pollution by carbon dioxide.

Catch crop: A rapidly growing plant that can be intercropped between rows of the main crop; often used as green manure.

Certificate of Registration: Shall mean the document issued by the Inspection and Certification Agency, declaring that the operator is licensed to use the certificate on specified products.

Certificate: Would mean a document issued by an accredited agency declaring that the operator is carrying out the activities or the stated products have been produced in accordance with the specified requirements in accordance with the National Standards for Organic Products.

Certification Mark: Shall mean certification programme's sign, symbol or logo which identifies the products as being certified according to the National Standards for Organic Products.

Certification Programme: Shall mean the system operated by an Inspection and Certification Agency in accordance with the criteria for carrying out certification of conformity as laid down herein.

Certification Transference: The formal recognition by an Inspection and Certification Agency of another Certification programme or Agency or projects or products certified by that programme or Agency, for the purpose of permitting its own certified operators to trade or process under the programme's own certification mark, the products which are certified by the other programme.

Certification: Shall be the procedure by which a written assurance is given by the Certification Agency that a clearly identified production or processing system has been methodically assessed and conforms to the specified requirements.

Certified organic agriculture: Certified organic agriculture refers to agricultural systems and products that have been managed and produced in accordance with specific standards or technical regulations and that have been inspected and approved by a certification body.

Certified organic farm: Any farm whose adherence to organic farming practices is certified against organic standards.

Certified organic food: Foods are produced according to organic agriculture standards. For crops, it means they were grown based on a system of farming that maintains and replenishes soil fertility and crop health without the use of conventional pesticides, artificial fertilizers, human waste, or sewage sludge, and that they were processed without ionizing radiation or food additives. For animals, it means they were reared without the routine use of antibiotics and without the use of growth hormones. Organic produce must not be genetically modified. Products usually are certified by a third-party certification body recognized at the international or national level, hence accountable in the case of fraud. Certification is made against the standards of the country where the product is sold. Certified organic food is recognized on the market by the organic label of the certification body.

Certified organic wild area: Refers to the organic certification of naturally grown plants in an approved and clearly defined collection area. The area itself is not certified. Plant products which grow in the wild can be certified as organic, but not every plant collected in its natural habitat can be considered as organic, as the natural habitat may be contaminated. Standards on organic wild collection require a clean collection area, sustainable collection and full traceability. The people who harvest, gather or wild craft shall not take any products at a rate that exceeds the sustainable yield of the harvested product, nor threaten the ecosystem stability or the existence of plant, fungal or animal species, including those not directly exploited.

Chain of Custody: All relevant steps in the production chain including growing, harvesting, processing, handling and related activities detailed in Section 4 of the accreditation criteria that have been inspected and certified, as appropriate.

Chain of custody: The concept that all relevant steps in the production chain including the growing, handling, processing and other processes have been inspected or certified as appropriate. In the traceability process, the chain of custody refers to the steps needed as a whole to trace a product from the source to the consumer.

Climate change adaptation: Adjustment in natural or human systems to a new or changing environment. Adaptation to climate change refers to adjustment in natural or human systems in response to actual or expected climatic stimuli or their effects, which moderates harm or exploits beneficial opportunities. Various types of adaptation can be distinguished, including anticipatory and reactive adaptation, private and public adaptation, and autonomous and planned adaptation.

Climate change mitigation: Intervention or policies to reduce the emissions or enhance the sinks of greenhouse gases. The current international legal mechanism for countries to reduce their emissions is the United Nations Framework Convention on Climate Change (UNFCCC).

Climate change: A change of climate which is attributed directly or indirectly to human activity that alters the composition of the global atmosphere and which is, in addition to natural climate variability, observed over comparable time periods.

Climate variability: Climate variability refers to variations in the mean state and other statistics (such as standard deviations, the occurrence of extremes, *etc.*) of the climate on all temporal and spatial scales beyond that of individual weather events.

Coexistence: Co-existence refers to the ability of farmers to make a practical choice between conventional, organic and genetically modified (GM) production, in compliance with the legal obligations for labelling and/or purity criteria. The possibility of adventitious presence of GM crops in non-GM crops cannot be dismissed and may have commercial implications for the farmers whose crops are affected. Consequently, suitable measures during cultivation, harvest, transport, storage, and processing may be necessary to ensure coexistence. The European Commission passed non-binding guidelines on coexistence in 2003. The differences in the level of implementation and in the practical form of national coexistence legislation are therefore great.

Companion planting: Crops that are planted close to one another to achieve some mutual benefit such as repelling insect pests or attracting beneficial insects, shade, wind protection, support, or nutrient enrichment.

Competent Authority: Shall mean the official government agency for accreditation.

CONTAMINATION: Pollution of organic product or land; or contact with any material that would render the product unsuitable for organic certification.

CONSULTANCY: Shall mean the advisory service for organic operations, independent from inspection and certification procedures.

Compost: A mixture of decaying organic matter, as from leaves and manure, used to improve soil structure and provide nutrients.

Conservation agriculture: Conservation agriculture aims to achieve sustainable and profitable agriculture and subsequently aims at improved livelihoods of farmers through the application of the three CA principles: minimal soil disturbance, permanent soil cover and crop rotations.

Conservation of natural resources: The protection, preservation, management, or restoration of wildlife and of natural resources such as forests, soil, and water. Conservation of natural resources is usually embraced in the broader concept of conserving the earth itself by protecting its capacity for self-renewal. It may be defined as the protection of natural resources and landscapes for later use.

Conservation tillage: It is a practice used in conventional agriculture to reduce the effects of tillage on soil erosion, however, it still depends on tillage as the structure-forming element in the soil.

Conservation: Includes protection, maintenance, rehabilitation, restoration and enhancement of populations and ecosystems. This implies sound biosphere management within given social and economic constraints, producing goods and services without depleting natural ecosystem diversity.

Contaminant: Any substance not intentionally added to food, which is present in such food as a result of production (including operations carried out in crop and animal husbandry), manufacture, processing, preparation, treatment, packing, packaging, transport or holding of such food or as a result of environmental contamination. The term includes chemical and biological substances not desirable in food but does not include insect fragments, rodent hairs and other extraneous matter.

Conventional agriculture: What is accepted as the norm and is the most dominant agricultural practice. Since World War II, (mainly in the industrialized world), conventional agriculture has become an industrialized form of farming characterized by mechanization, monocultures, and the use of synthetic inputs such as chemical fertilizers, pesticides and genetically modified organisms (GMOs), with an emphasis on maximizing productivity and profitability and treating the farm produce as a commodity. In large parts of the developing world, agriculture is still "traditional", ranging from well-managed polycultures to extensive and eroding pastures.

Conventional: Farming systems dependent on input of artificial fertilizers and/ or chemicals and pesticides or which are not in conformity with the basic standards of organic production.

Conversion period: The conversion period is the time between the start of organic management and the certification of crops or animal husbandry as organic. It is the time taken to clean-up chemical residues, if any, left behind in the soil by previous agricultural techniques and re-establish the ecological balance (2-3 years) necessary for soil fertility and pest-predator balance. The start of the conversion period shall be calculated from the date of application to the certification body or, alternatively, from the date of the last application of unapproved inputs, provided that the operator can demonstrate that the requirements of the full standard have been met for at least the minimum period of 12 months prior to pastures, meadows and products harvested therefrom, being considered organic. In the case of perennials (excluding pastures and meadows) a period of at least 18 months prior to harvest shall be required. The conversion period for dairy products is a minimum of 90 days and for eggs 42 days.

Conversion: The process of changing an agricultural farm from conventional to organic farm. This is also called transition.

Corporate social responsibility: A set of management practices in businesses that aim at minimizing the negative impacts of their operations on society and at maximizing the positive impacts. A concept, whereby companies integrate social and environmental concerns in their business operations and in their interaction with their stakeholders on a voluntary basis. CSR covers social and environmental issues, in spite of the English term corporate social responsibility. An important aspect of CSR is how enterprises interact with their internal and external stakeholders: employees, customers, neighbours, non-governmental organizations, public authorities, *etc*

Corrective action: Action to eliminate the cause of a potential nonconformity or other undesirable situation. In organic agriculture, corrective actions are verified by the certification body, with a view to protecting the organic claim.

Cover crop: A crop grown to prevent soil erosion by covering the soil with living vegetation and roots that hold on to the soil. Cover crops are also grown to help maintain soil organic matter and increase nitrogen availability (green manure crop), and to "hold on" to excess nutrients (a catch crop) still in the soil, following an economic crop. Other benefits of cover crops include weed suppression and attraction of beneficial insects.

Crop ecology: Crop relation or interactions with its biotic (*e.g.* pests) and abiotic environment (*e.g.* soil) and which determines crop growth. Crop ecology evolved at the end of the 1920s, focusing on the study of the physical and environmental conditions in which crops were grown in order to identify the best places where to cultivate them.

Crop rotation: The practice of alternating the species or families of annual and/or biannual crops grown on a specific field in a planned pattern or sequence so as to break weed, pest and disease cycles and to maintain or improve soil fertility and organic matter content.

Cross-fertilization: Fertilization by pollen from another plant. The transfer of pollen from the flowers of one plant to the stigma of another plant. It may or may not lead to fertilization.

Declaration of Interest: Declaration of no personal/commercial conflict of interest by all concerned involved in the process of inspection and certification.

Dehorning: Dehorning is the process of removing or stopping the growth of the horns of livestock. Cattle, sheep, and goats are dehorned for economic and safety reasons.

Disinfectant: A product that minimizes by physical or accepted chemical means, the number of microorganisms in the environment, to a level that does not compromise food safety and suitability.

Diversified production: Different mix of crops, trees, animals, fish to ensure a variety of food, fodder and fibre sources and complimentary use of natural resources. It also brings more ecosystem stability. Mixed cropping is a system of sowing two or three crops together on the same land, one being the main crop and the others the subsidiaries.

Diversity: Species richness of a community or area, measured in terms of the number of different plant and animal species (often called species richness) it contains. However, the community characteristics are better assessed by the relative abundance of the species present. Diversity in ecosystems is usually equated with stability due to the climax community.

Drought-resistant crop: Crops that can dwell in conditions of water shortage. Drought-tolerant crops are selected for their resilience to drought. With the intensification of droughts caused by climate change, breeding drought-tolerant crops is important for food security. Local crops provide the gene pool necessary to select the most adapted varieties for farming under drought conditions.

Dung: Organic material that is used to fertilize the land, usually consisting of the faeces and urine of domestic livestock, with or without litter such as straw, hay, or bedding. Some countries also use human excrement ("night soil"). Though livestock manure is less rich in nitrogen, phosphorus, and potash than synthetic fertilizers and therefore must be applied in much greater quantities, it is rich in organic matter, or humus, and thus increases soil fertility and improves the capacity of soil to absorb and store water, thereby preventing erosion. Because manure must be carefully stored and spread in order to derive the most benefit, some farmers decline to expend the necessary time and effort. Manufactured chemical fertilizers, though more concentrated and efficient, are also more costly and more likely to cause excess runoff and pollution.

Eco-labelling: Voluntary method of environmental performance certification and labelling. An "eco-label" is a label which identifies the overall environmental preference of a product or service based on life cycle considerations. In contrast to "green" symbols or claim statements developed by manufacturers and service providers, an eco-label is awarded by an impartial third-party in relation to certain products or services that are independently determined to meet environmental leadership criteria.

Ecological agriculture: Ecological agriculture is a management system that enhances natural regenerative processes and stabilizes interactions within local agro-ecosystems. Ecological agriculture includes organic agriculture as well as other ecological approaches to farming that allow the use of synthetic inputs.

Ecological awareness: Ecological awareness arises when people or more specifically consumers are concerned and aware of ecological issues and this can be a first step in the direction of changing attitudes towards the products they want to buy and/or their behaviours to respect the environment.

Ecological balance: A state of dynamic equilibrium within a community of organisms in which genetic, species and ecosystem diversity remain relatively stable, subject to gradual changes through natural succession.

Ecological footprint: The ecological footprint is a measure of human demand on the Earth's ecosystems; it compares human demand with planet Earth's ecological capacity to regenerate it. It represents the amount of biologically productive land and sea area needed to regenerate the resources a human population consumes and to absorb and render harmless the corresponding waste, given prevailing technology and resource management practice. Using this assessment, it is possible to estimate how many planet Earths it would take to support humanity if everybody lived a given lifestyle. While the ecological footprint term is widely used, methods of measurement vary. But calculation standards are now emerging to make results more comparable and consistent.

Ecological intensification: Maximization of primary production per unit area without compromising the ability of the system to sustain its productive capacity. This entails management practices that optimize nutrient and energy flows and use local resources, including horizontal combinations (such as multiple cropping systems or polycultures); vertical combinations (such as agroforestry); spatial integration (such as crop-livestock or crop-fish systems); and temporal combinations (rotations).

Ecological management: The management of human activities so that ecosystems, their structure, function, composition, and the physical, chemical, and biological processes that shape them continue to renew themselves. Sometimes called an ecological approach to management.

Ecological quality: Ecological quality is defined as the overall expression of the structure and function of an ecosystem. It is expressed by a number of ecological quality elements or variables, reflecting the different parts of the ecosystem, to which conservation and use objectives or targets can be set.

Ecological resilience: Capacity of a natural ecosystem to recover from disturbance.

Ecological tourism: Travel to a pristine natural area that appeals to environmentally conscious individuals. An integral part of ecological tourism is the promotion of recycling, energy efficiency and water conservation in order to minimize their impact and conserve the environment.

Evaluation: Shall be the process of systematic examination of the performance of an Inspection and Certification Agency to meet the specific requirements under the National Accreditation Programme.

Farm Unit: An agricultural farm, area or production unit managed organically, by a farmer or a group of farmers.

Food Additive: Food additive is an external permissible ingredient added to improve the keeping quality, consistency, colour and other physico-chemical, sensory properties, wholesomeness and safety of food.

Genetic Diversity: Genetic diversity means the variability among living organisms from agricultural, forest and aquatic ecosystem. This includes diversity within species and between species.

Genetically Modified Organism: A genetically modified/engineered organism means an organism in which the genetic material has been changed through modern biotechnology in a way that does not occur naturally by multiplication and/or natural recombination. For instance, a plant may be given fish genetic material that increases its resistance to frost. Another example would be an animal that has been modified with genes that give it the ability to secrete a human protein.

Genetically Modified Organism-Free Region: Zones, regions, provinces, even whole countries, where local/regional governments declared that genetically modified (GM) crops cannot be planted in order to protect conventional and organic crops, as well as wildlife, from potential GM contamination.

Global Warming Potential: The global warming potential (GWP) in agriculture is measured by the quantity of greenhouse gases emissions (expressed in CO_2 equivalent) that an activity is likely to produce both in GHG per hectare and per tonne of commodity. GWP is a measure of how much a given mass of greenhouse gas is estimated to contribute to global warming. It is a relative scale which compares the gas in question to that of the same mass of carbon dioxide.

Green Manuring: Green manuring refers to a cover crop grown to help maintain soil organic matter and increase nitrogen availability. Legumes are often used because they have rhizobial bacteria living in their root nodules that are able to fix nitrogen from the air and add it to the soil. Green manure is incorporated into the soil for the purpose of soil improvement. May include spontaneous crops, plants or weeds.

Greenhouse Gas: Greenhouse gases are those gaseous constituents of the atmosphere, both natural and anthropogenic that absorb and emit radiation at specific wavelengths within the spectrum of infrared radiation emitted by the Earth's surface, the atmosphere, and clouds. This property causes the greenhouse effect. Water vapour (H_2O), carbon dioxide (CO_2), nitrous oxide (N_2O), methane (CH_4), and ozone (O_3) are the primary greenhouse gases in the Earth's atmosphere. Moreover there are a number of entirely human-made greenhouse gases in the atmosphere, such as the halocarbons and other chlorine- and bromine-containing substances, dealt with under the Montreal Protocol. Besides CO_2, NvO, and CH_4, the Kyoto Protocol deals with the greenhouse gases sulfur hexafluoride (SF_6), hydrofluorocarbons (HFCs), and perfluorocarbons (PFCs).

GROUP CERTIFICATION: Certification of an organized group of producers, processors and exporters with similar farming and production systems and which are in geographical proximity.

Grower Group: Grower groups are an organized group of producers with similar farming and production systems, working according to a common marketing objective.

Growth Promoter: Growth promoters are synthetic substances that are included to the feed in order to maximise growth of animals; when applied to a plant, they promote, inhibit or otherwise modify the growth of a plant. These substances are forbidden in organic agriculture.

Guidelines For Organic Production and Processing: Standards for organic production and processing established by the Accreditation Agencies for specific crops in accordance with the National Standards for Organic Products.

Hazard Analysis Critical Control Point (HACCP): A systematic process that identifies food safety hazards, critical control points, critical limits, corrective actions and documentation and integrates monitoring procedures to ensure food safety.

Humus: Decomposed, dark brown and amorphous organic matter of soils, having lost all trace of the structure and composition of the vegetable and animal matter from which it was derived. Humus hence refers to any organic matter that has reached a point of stability and which is used in agriculture to amend soil.

Ingredient: Shall mean any substance, including a food additive, used in the manufacture or preparation of a food and present in the final product although possibly in a modified form.

In-situ Conservation: The conservation of ecosystems and natural habitats and the maintenance and recovery of viable populations of species in their natural surroundings and, in the case of domesticated or cultivated species, in the surroundings where they have developed their distinctive properties.

Inspection Agency: Shall mean the agency that performs inspection services as per the National Accreditation Policy and Programme.

Inspection: Shall include the site visit to verify that the performance of an operation is in accordance with the production or processing standards.

Inspector: Shall be the person appointed by the Inspection and Certification Agency to undertake the inspection of an operator.

Integrated Natural Resources Management: INRM is the term used by the Consultative Group for International Agricultural Research (CGIAR) for research aiming at improving livelihoods, agroecosystem resilience, agricultural productivity and environmental services. The approach seeks to integrate broad-based management of the land, water, forest and biological resource base (including genes) needed to sustain agricultural productivity and avert degradation of potential productivity.

Integrated Pest Management: Integrated pest management (IPM) means the careful consideration of all available pest control techniques and subsequent integration of appropriate measures that discourage the development of pest populations and keeps pesticides and other interventions to levels that are economically justified and reduce or minimize risks to human health and the environment. IPM emphasizes the growth of a healthy crop with the least possible disruption to agro-ecosystems and encourages natural pest control mechanisms.

Integrated production: System that mixes plant, livestock, trees and/or fish, produced contemporarily. Emphasis is placed on a holistic systems approach involving the entire farm as the basic unit and on balanced nutrient cycles. Biological, technical and chemical methods are balanced carefully taking into account the protection of the environment, profitability and social requirements.

Intercropping: Growing two or more crops as a mixture in the same field at the same time. Intercropping can be one way of adding diversity to a crop system.

Internal Control System: A documented quality assurance system that allows the external certification body to delegate the inspection of individual group members to a body identified from within the operators of the group.

Internal Review: Shall mean an assessment of the objectives and performance of a programme by the Certification or the Accreditation Agency itself.

International organic standard: The Codex Alimentarius and IFOAM guidelines are minimum standards for organic agriculture, intended to guide, respectively, governments and private certification bodies in standard setting. As such, they can be considered as standards for standards. Governments can use these texts to develop national organic agriculture programmes, which are often more detailed as they respond to specific country needs.

In-transition: Period of conversion to organic, from a previous management system, be it industrial or traditional. Crops grown on land in transition to organic (during the first two to three years after switching from conventional farming) cannot be labelled as organic.

Irradiation: High energy emissions from substances for the purpose of controlling microbial, pathogens, parasites and pests in food, preserving the food or inhibiting physiological processes such as sprouting or ripening.

Labelling: Means any written, printed or graphic matter that is present on the label, accompanies the food, or is displayed near the food, including that for the purpose of promoting its sale or disposal.

Land conversion: Converting an area to another use such as converting forest area or wetlands into agricultural land or urban area. Conversion is often confused with clear-cut. An area that is clear-cut remains forested.

Legume-based organic rotation: A traditional component of crop rotation is the replenishment of nitrogen through the use of legumes in sequence with other crops. Legume-based rotations increase soil fertility by fixing nitrogen.

Licence: Shall be the Accreditation contract that grants a certifier the rights associated with its accredited status in line with the National Program for Organic Production.

Livestock: Shall mean any domestic or domesticated animal including bovine (including buffalo and bison), bovine, porcine, caprine, equine, poultry and bees raised for food or in the production of food. The products obtained by hunting or fishing of wild animals shall not be considered as part of this definition.

Micro-organism: An organism of microscopic or submicroscopic size, especially a bacterium or protozoan.

Mineral fertilizer: Fertlizers manufactured by chemical and industrial processes. May include products not found in nature, or simulation of products from natural sources (but not extracted from natural raw materials). It refers to agricultural substances produced through chemical processes, including nitrogen-fertilizers.

Minimum tillage: Minimum tillage is a tillage method that does not turn the soil over, with a view to maintain biodiversity structure.

Mulching: A protective covering, usually of organic matter such as leaves, straw, or peat, placed around plants to prevent the evaporation of moisture, the freezing of roots, and the growth of weeds.

Mycorrhiza: Fungi that form an association with, or have a symbiotic relationship with roots of more developed plants. Mycorrhiza improve soil fertility as they improve the mineral absorption capabilities of the plant roots and consequently, resistance to diseases.

Mycotoxin: Toxic substance of fungal origin (*e.g.* aflatoxin) that proliferates on crops at specific level of moisture, temperature and oxygen in air.

Natural resources: Any portion of the natural environment, such as air, water, soil, botanical and zoological resources and minerals. A renewable resource can potentially last indefinitely (provided stocks are not overexploited) without reducing the available supply because it is replaced through natural processes (either because it recycles rapidly, as water does, or because it is alive and can propagate itself or be propagated, as organisms and ecosystems do). Non-renewable resources (such as coal and oil) may eventually be replaced by natural processes, but these processes occur over long periods of geologic time rather than within the time-frame of current generations, and their consumption necessarily involves their depletion.

Non-certified organic agriculture: In many developing countries, there are agricultural systems that fully meet the requirements of organic agriculture but which are not certified. Non-certified organic agriculture refers to organic agricultural practices by intent and not by default; this excludes non-sustainable systems which do not use synthetic inputs but which degrade soils due to lack of soil building practices. It is difficult to quantify the extent of these agricultural systems as they exist outside the certification and formal market systems. The produce of these systems is usually consumed by households or sold locally (*e.g.* urban and village markets) at the same price as their conventional counterparts. In developed countries, non-certified organic food is often sold directly to consumers through local community support programmes such as box schemes, farmers markets and at the farm gate.

Non-certified organic farmer: There are organic farmers for whom certification does not have any advantages: this is true for farmers who practice subsistence farming, basically catering for the food security of their families or their community. It is also true for farmers who want to sell their produce as organic, where a demand for organic products does not exist in their region or where the intermediary or processor does not want to handle organic products. There are also farmers that reject certification on principal or economic grounds.

Organic agriculture action plan: An official document specifying the policy objectives, strategy, actions and programmes necessary to support the development of the organic sector.

Organic agriculture market: Organic markets are growing but reactive, driven by food safety concerns and to a lesser extent, by environmental awareness. They often establish producer consumer groups to provide direct food marketing through such activities as farmers' markets or home deliveries to subscribed customers, which increases profits.

Organic Agriculture Principle: The General Assembly of the International Federation of Organic Agriculture Movements (IFOAM) approved 4 principles of organic agriculture upon which organic agriculture is based: the principle of health; the principle of ecology; the principle of fairness; the principle of care. Principles apply to agriculture in the broadest sense, including the way people tend soils, water, plants and animals in order to produce, prepare and distribute goods. They concern the way people interact with living landscapes, relate to one another and shape the legacy of future generations. Each principle is followed by an action oriented explanation.

Organic Agriculture Standard: Organic standards have long been used to create an agreement within organic agriculture about what an "organic" claim on a product means, and to some extent, to inform consumers. It includes recommended and prohibited practices and substances as well as guarantee requirements. Regional groups of organic farmers and their supporters began developing organic standards as early as in the 1940s. Currently, there are over 450 private organic standards worldwide; and in addition, organic standards have been codified in the technical regulations of more than 60 governments.

Organic agriculture systems contribute to reduced consumption of fossil-fuel energy (by foregoing use of nitrogen fertilizers), reduced greenhouse gas emissions, reduced soil erosion and increased carbon stocks, especially in already degraded soils. Greenhouse warming potential in organic systems is 29 to 37 per cent lower, on a per hectare basis, because of omission of synthetic fertilizers and pesticides as well as less use of high energy feed. Methane emissions of organic rice and ruminants are equal to conventional systems but the increased longevity of organic cattle is favourable on methane emissions. Carbon sequestration efficiency of organic systems in temperate climates is almost double as compared to conventional soils, mainly due to use of grass clovers for feed and of cover crops in organic rotations.

Organic Agriculture: Organic agriculture is a holistic production management system which promotes and enhances agroecosystem health, including biodiversity, biological cycles, and soil biological activity. It emphasizes the use of management practices in preference to the use of off-farm inputs, taking into account that regional conditions require locally adapted systems. This is accomplished by using, where possible, cultural, biological and mechanical methods, as opposed to using synthetic materials, to fulfill any specific function within the system. Organic agritourism is committed to organic farming as well as to energy saving; reduction of air emissions; improvement of the use of water and reduction and recycling of wastes. It is important that the organic agritourism establishes a connection with the active elements and actors of the territory so as to create a real network connecting territory, culture and traditions. Organic farms can be organic agritourism offering to tourists a variety of activities which: contribute to the conservation of nature, biodiversity and cultural heritage; respect the integrity of ecosystems and habitats; benefit local communities; use environmentally-sound technologies; minimize the production of waste and encourage recycling; encourage the use of public and non-motorised transport.

Organic Aquaculture: Aquatic species produced according to organic standards. Most reported certified organic aquaculture products produced in Europe use marine and brackish waters, a largely untapped resource, thus preserving freshwater supplies for human consumption and agriculture. Aquaculture also covers organic aquatic plants for either direct human consumption or for use as feed inputs for animal husbandry, including for the organic aquaculture sector.

Organic Breeding: According to IFOAM, the general principle for organic breeding is that breeds are adapted to local conditions. IFOAM recommends that breeding goals should encourage and maintain the good health and welfare of the animals consistent with their natural behaviour. Breeding practices should include methods that are not capital intensive methods or depend on high technologies invasive to natural behaviour. Animals should be bred by natural reproduction techniques. Standards should require that breeding systems shall be based on breeds that can reproduce successfully under natural conditions without human involvement. Artificial insemination is permitted. Hormones are prohibited to induce ovulation and birth unless applied to individual animals for medical reasons and under veterinary supervision.

Organic Commodity: According to the general definition of a commodity: a physical substance, such as food, grains, and metals, which is interchangeable with another product of the same type, and which investors buy or sell, usually through contracts. The price of the commodity is subject to supply and demand. The concept of commodity entered the field of organic agriculture since the organic sector is linked to trade and it has become a huge market (46 billion US$ in 2008).

Organic Community: The "organic" community gathers all the relevant actors which operate in the sector of the organic agriculture, such as relevant policy and standard-setting institutions, as well as individuals and groups involved with production, processing, certifying, commercializing and consuming organic good and services.

Organic Compound: In physics, a material that contains carbon and hydrogen and usually other elements such as nitrogen, sulphur and oxygen. Organic compounds can be found in nature or they can be synthesized in the laboratory. An organic substance is not the same as a "natural" substance. A natural material means that it is essentially the same as it was found in nature, but "organic" means that it is carbon-based.

Organic Conversion: Process of change into an organic agricultural system from a different management system, industrial or traditional or integrated it may be.

Organic Ecosystem Management: Management that includes principles, recommendations and requirements for maintaining and improving: landscape and biodiversity quality; soil and water quality; prohibition on clearing primary ecosystems; exclusion of genetic engineering from organic production and processing; and prevention of degradation of common/public lands when harvesting or gathering wild products.

Organic Farm: Any farm which uses organic farming practices. Organic farming is more than agricultural production without the use of synthetic chemicals or genetically modified organisms, growth regulators, and livestock feed additives. Organic farming emphasises a holistic farm management approach, where rotations and animals play an integral role in the system.

Organic Fertilization: The use of natural organic fertilizer that helps to provide all the nutrients required by the plants and increase the quality of the soil with a natural microorganism environment.

Organic Food Processing: Organic food is to be processed by biological, mechanical and physical methods in a way that maintain the vital quality of each ingredient, the finished product and nutritional value. Processors should choose methods that limit the number and quantity of non-organic additive and processing aids. Any additives, processing aids or other material that chemically react with or modify organic food shall be restricted. Irradiation is not permitted. Filtration equipment shall not contain asbestos, or utilize techniques or substances that may negatively affect the product. The following conditions of storage are permitted: controlled atmosphere, temperature control; drying and humidity regulation. Use of approved processing aids includes: drying with ascorbic acid, citric acid, tartaric acid and salt; blanching with high temperatures to destroy microorganisms; pasteurizing to destroy micro-organisms that could contaminate the product after blanching; and with heat treatments that conserve products by destroying or inactivating enzymes and killing micro-organisms.

Organic Manure: Organic manure covers manures made from cattle dung, excreta of other animals, rural and urban composts, other animal wastes, crop residues and last but not the least green manures. Organic manure is time tested materials for improving the fertility and productivity of soils.

Organic Matter: Plant and animal residues at various stages of decomposition, cells and tissues of soil organisms, and substances synthesized by the soil population.

Organic Pasture: Organic pasture is the main activity which can benefit from the conservation of biodiversity. Organic pasture management reflects a synthesis of crop and livestock production principles that works from the soil up to promote an interdependent community of plants and ruminants. Organically managed pasture should produce the quantity and quality of edible plants suitable to the species, stage of production, and number of animals. Access to pasture assures a relationship between the animal and land that satisfies both organic principles and international standards for organic livestock.

Organic Pest Management: Today, insect pest management in organic agriculture involves the adoption of scientifically based and ecologically sound strategies as specified by international and national organic production standards. These include a ban on synthetic insecticides and, more recently, on genetically modified organisms (GMOs). Pest management in organic systems differ from conventional agriculture conceptually in that indirect or preventative measures form the foundation of the system, while direct or reactive control methods are rare and must comply with organic production standards. Pest control in organic agriculture begins by making sensible choices, such as growing crops that are naturally resistant to diseases and pests, or choosing sowing times that prevent pest and disease outbreaks. Substituting synthetic pesticides with biological pest control substances is part of the strategy during conversion but is not economically efficient neither desirable once the pest-predator balance is re-established in the system.

Organic Post-Harvest Handling: Includes all stages of production immediately following harvest, including cleaning, cooling, sorting, storing and packing. Many post-production operations for organic produce are identical to non-organic production. Where there are particular restrictions or considerations they are identified. The draft IFOAM Basic Standards 2002 state that: 'Handlers and processors should handle and process organic products separately in both time and place from non-organic products. Handlers and processors should identify and avoid pollution and potential contamination sources'. Likewise the Codex Alimentarius (Annex 1B) required the maintenance of organic product integrity and protection against contamination.

Organic Production Unit: Shall mean a unit/holding or stock farm complying with the rules of NPOP regulations.

Organic Seeds And Planting Material: Seed and planting material produced under certified organic system.

Organic soil fertility management: Organic soil fertility management is guided by the philosophy of "feed the soil to feed the plant". This basic precept is implemented through a series of practices designed to increase soil organic matter, biological activity, and nutrient availability.

Organic urban garden: Refers usually to private gardens situated in the city area, farmed by their owners following organic agriculture principles.

Organic yield: Refers to the accumulated volume or biomass remaining from gross production in an organic crop, livestock and farmed fish systems. Organic yields are lower when compared to high external input systems and higher when compared to low external input systems. Comparing crop-specific yields, however, does not account for the whole biomass production (including crops, stems and roots), of the rotation period and of the whole farm production.

Organically-Produced Feeding Stuffs/Feed Materials: Shall mean feeding stuffs/ feed material produced in accordance with the rules of production laid down in NPOP regulations.

Organoleptic: Refers to any sensory properties of a food or other products, including taste, colour, odour and texture.

Package of Practices: Guidelines for organic production and processing established by the Accreditation Agencies for specific crops, specific to the region.

Parallel Production: Shall mean any production where the same unit is growing, breeding, handling or processing the same products both in a certified organic and a non-certified organic system. Similarly a situation with "organic" and "in conversion" production of the same product is also parallel production.

Part Conversion: Shall be when part of a conventional farm or unit has already been converted to organic production or processing and a part is in the process of conversion.

Plant Protection Product: Plant protection product means any substance intended for preventing, destroying, attracting, repelling, or controlling any pest or disease including unwanted species of plants or animals during the production, storage, transport, distribution and processing of food, agricultural commodities, or animal feeds.

Plant Protection Product: Shall mean any substance intended for preventing, destroying, attracting, repelling, or controlling any pest or disease including unwanted species of plants or animals during the production, storage, transport, distribution and processing of food, agricultural commodities, or animal feeds.

Predation: Predation is the transfer of energy whereby one organism feeds on another organism as well as complex interactions among predators-prey populations. If a portion of the prey is not available because of environmental discontinuities (a typical case in agriculture), the self-regulating balance will be damped. Inter-specific competition keeps more pests in check than we ever could by using pesticides. Ecosystem stability (or instability) depends on the results of the competition between different species for food and space. Predation ameliorates the intensity of competition for space and increases species diversity.

Preparation: Shall mean the operations of slaughtering, processing, preserving and packaging of agricultural and animal products and also alterations made to the labeling concerning the presentation of the organic production method.

Quality management system: Quality management system refers to a system to direct and control an organization with regard to quality.

Rhizome: In botany, a rhizome is a horizontal stem of a plant that is usually found underground, often sending out roots and shoots from its nodes. Plants with underground rhizomes include ginger, hops, and turmeric, significant for their medicinal properties, and the weeds Johnson grass, bermuda grass, and purple nut sedge. Some plants have rhizomes that grow above ground or that sit at the soil surface, including some Iris species, and ferns, whose spreading stems are rhizomes. Rhizomes may also be referred to as creeping rootstalks, or rootstocks.

Soil biodiversity: Soil is one of the most diverse habitats on earth and contains one of the most diverse assemblages of living organisms. Nowhere in nature are species so densely packed as in soil communities. For example a single gram of soil may contain millions of individuals and several thousands species of bacteria. Soil biota includes micro-organisms (bacteria, fungi, *etc.*), microfauna (protozoa, nematodes, *etc.*), meso-fauna (acari, springtails, *etc.*) and macrofauna (insects, earthworms, *etc.*). It also includes the roots that grow in the soil and interact with other species above and below ground.

Soil health: The terms soil quality (favoured by scientists) and soil health (favoured by farmers) tend to be used interchangeably. Characterization of soil quality by scientists focuses on analytical/quantitative properties of soil with a separately defined quantitative link to the functions of soil quality. Characterization of soil health by farmers focuses on descriptive/qualitative properties of soil with a direct value judgement (unhealthy to healthy) integrated into the options for a given property; in addition, interwoven into the properties of soil per se are value based descriptive properties of plant, water, air, and animal/human systems considered by farmers to be an integral part of soil health characterization.

Soil organic carbon: Soil organic carbon (SOC) refers to the carbon held within the soil and is expressed as a percentage by weight (gC/Kg soil). Climatic shifts in temperature and precipitation have a major influence on the decomposition and amount of SOC stored within an ecosystem and that released into the atmosphere. Globally, the amount of carbon stored in soils is twice the amount that is stored in all terrestrial plants. Soil organic carbon (SOC) is essential for maintaining fertility, water retention, and plant production in terrestrial ecosystems. The amount of SOC stored within an ecosystem, is dependent on the quantity and quality of organic matter returned to the soil matrix, the soils ability to retain organic carbon (a function of texture and caption exchange capacity), and biotic influences of both temperature and precipitation. The global decline in SOC as a result of deforestation, shifting cultivation and arable cropping have made significant contributions to increased levels of atmospheric carbon dioxide (CO2).

Soil Organic Matter: Soil organic matter (SOM) is defined as all organic materials found in soils irrespective of origin or state of decomposition. It can be divided into three general pools: living biomass of micro-organisms, fresh and partially decomposed residues, and the well-decomposed and highly stable organic material, or humus.

Soil Resilience: Soil have an inherent ability to restore their life support processes, provided that the disturbance created especially by human activities is not too drastic, and sufficient time is allowed for the life support processes to restore themselves. This intrinsic soil productivity regeneration ability is called "resilience".

Soil Retention: The spaces that exist between soil particles, called pores, provide for the passage and/or retention of gases and moisture within the soil profile. The soil's ability to retain water is strongly related to particle size; water molecules hold more tightly to the fine particles of a clay soil than to coarser particles of a sandy soil, so clays generally retain more water. Conversely, sands provide easier passage or transmission of water through the profile. Clay type, organic content and soil structure also influence soil water retention. Soil water retention is essential to life. It provides an ongoing supply of water to plants between periods of replenishment (infiltration) so as to allow their continued growth and survival.

Soil Stability: Soil stability depends on soil's shear strength, its compressibility and its tendency to absorb water. Farming practices that preserve soil fertility and maintain, or even increase, organic matter in soils can reduce the negative effects of drought while increasing primary crop productivity.

Some physical alterations, such as beak trimming, are prohibited by some certification bodies, while allowed by others if the practice is done to improve or maintain the health and safety of the animal. Dehorning is allowed in organic agriculture but not in biodynamic agriculture.

Spray Preparation: Substances used for plant protection and plant growth regulator in biodynamic agriculture. The biodynamic preparations consist of recycled mineral, plant or animal manure extracts that are fermented over time and added in homeopathic or very dilute quantities, to compost piles, manure and slurry, which are then applied to the soil or sprayed directly onto plants.

Surveillance: The measures undertaken to provide monitoring of an operator's/ certification body's compliance with the standards/criteria for meeting the certification/accreditation requirements.

Traditional Agriculture: Traditional agriculture, is an indigenous form of farming, result of the co-evolution of local social and environmental systems and that exhibit a high level of ecological rationale expressed through the intensive use of local knowledge and natural resources, including the management of agrobiodiversity in the form of diversified agricultural systems.

Transaction/Import Certificate: Document issued by a certification body declaring that the specified lot or consignment of goods is derived from production and/or processing system that has been certified.

Transgenic Organism: An individual in which a transgene has been integrated into its genome. In transgenic eukaryotes, the transgene must be transmitted through meiosis to allow its inheritance by the offspring.

Urban and Peri-Urban Organic Agriculture: Urban (and peri-urban) agriculture is the practice of agriculture (including crops, livestock, fisheries, and forestry activities) within or surrounding the boundaries of cities. The land used may be private residential land (use of private pieces of land, balconies, walls or building roofs), public roadside land or river banks and peri-urban open fields. Urban farming is practiced for income-earning or food-producing activities. It contributes to food security and food safety in two ways: first it increases the amount of food available to people living in cities, and second it allows fresh vegetables and fruits to be made available to urban consumers. Organic agriculture is practiced in urban centres, including backyards and public space, and their periphery. In the Pro-huerta initiative in Argentina, for instance, urban agriculture provides food production and self-employment but also helps to 'create an improved microclimate and conserve soils, to minimize waste in cities and to improve nutrient recycle, and to improve water management, biodiversity, the O_2–CO_2 balance, and the environmental awareness of city inhabitants.

Utilized Agricultural Area: All the area of arable land, permanent meadow and pasture, and land devoted to permanent crops and kitchen gardens.

Vermicomposting: Vermicompost (or worm compost) the process of using earthworms to breakdown kitchen and garden waste, to create a faster than normal composting. Compared to ordinary soil, the earthworm castings (the material produced from the digestive tracts of worms) contain five times more nitrogen, seven times more phosphorus and 11 times more potassium. They are rich in humic acids and improve the structure of the soil. The earthworm most often to be found in the compost heap is Brandling Worms (*Eisenia foetida*), or Redworms (*Lumbricus rubellus*). This species is only rarely found in soil and is adapted to the special conditions in rotting vegetation, compost and manure piles. Earthworms are available from mail-order suppliers, or from angling shops where they are sold as bait. Small scale vermicomposting is well suited to turn kitchen wastes into high-quality soil where space is limited. In addition to worms, a healthy vermicomposting system hosts many other organisms such as insects, moulds, and bacteria. Though these all play a role in the composting process, the earthworm is the major catalyst for the composting process.

Vermiculture: The activity of growing and multiplication of earthworms. Usually done to either increase earthworm population in soil, for preparing vermicompost or for sale of worms to vermicomposting enterprises.

Zero tillage: Zero tillage is the simple technique of drilling seed into the soil with little or no prior land preparation. Zero tillage is a technical component used in conservation agriculture, but not everyone carrying out zero tillage is practising conservation agriculture.

References

A.K. Yadav, *Training Manual – Certification and Inspection Systems in Organic Farming in India*, http://ncof.dacnet.nic.in/Training_manuals/Training_manuals_in_English/Cert_and_Inspection_manual.pdf.

Akhurst, R. and K. Smith, 2002. Regulation and safety. In: Gaugler, R. (Ed.), *Entomopathogenic Nematology.* CABI, New York, NY, pp. 311-332.

Amir Cheraghi, Behzad Habibpour and Mohammad Saied Mossadegh, (2013). Application of Bait Treated with the Entomopathogenic Fungus Metarhizium anisopliae (Metsch.) Sorokin for the Control of Microcerotermes diversus Silv. Psyche Volume 2013 (2013), Article ID 865102, 5 pages http://dx.doi.org/10.1155/2013/865102

Anand, R. and B.N. Tiwary. (2009). Pathogenicity of entomopathogenic fungi to eggs and larvae of *Spodoptera litura*, the common cutworm. *Biocontrol Sci. Technol.*, 19(9): 919-929.

Ann, E. Hajek, (2012). An Introduction to Biological Control. Cambridge University Press, 2013.

Annie Gunn, *Organic Labelling Legal Requirements*, https://legalvision.com.au/organic-labelling-legal-requirements/, August 26, 2016.

Anthony Ditro, 2013. Florida predatory stink bug, *Euthyrhynchus floridanus* (Linnaeus). http://entnemdept.ifas.ufl.edu.

APEDA. Accredited inspection and certification bodies. In National program for organic agriculture (NPOP)– standards, 2006.

Aronstein, K., Oppert, B. and Lorenzen, M.D. (2011) RNAi in agriculturally-important arthropods. RNA Processing (ed. P. Grabowski), pp. 157–180. *InTech. Rijeka, Croatia.*

B. Sen, T.S. Chandra, 2007. Chemolytic and solid-state spectroscopic evaluation of organic matter transformation during vermicomposting of sugar industry wastes, *Bioresour. Technol.* 98, 1680–1683.

Bailey, A. S., Bertaglia, M., Fraser I. M. and Sharma A. Douarin, 2009. Integrated Pest Management portfolios in UK arable farming: results of farmers survey. *Pest Management Science* 65, 1030-1039.

Bajsa, O., Nair, J., Mathew, K., Ho, G.E., 2003. Vermiculture as a tool for domestic wastewater management. *Water Science and Technology* 48 (11–12), 125–132.

Barrett, H.R., Browne, A.W., Harris, P.J.C. and Cadoret, K., 2002. Organic certification and the UK market: organic imports from developing countries. *Food policy,* 27(4), pp.301-318.

Biological Control Editorial Board. Elsevier Journal website. Copyright © 2013 Elsevier B.V. Karen A. O'Hanlon, Kamilla Knorr, Lise Nistrup J rgensen, Mogens Nicolaisen, Birte Boelt, 2012. Exploring the potential of symbiotic fungal endophytes in cereal disease suppression. *Biological Control* 63 (2012) 69–78 Review.

Biological Control of Weeds: A Handbook for Practitioners and Students (1992), K. Harley and I.W. Forno. Inkata Press, Melbourne, Australia. 74 p.

Biological Control of Weeds: Theory and Practical Application (1997), M. Julien and G. White. ACIAR Monograph No. 49, Australian Centre for International Ag. Research, Canberra, Australia. 192 p.

Biologically Based Technologies for Pest Control (1995), a 216-page report by the Office of Technology Assessment. Available for $14 from the Superintendent of Documents, P.O. Box 371954, Pittsburgh, PA 15250-7954. (Phone: 202-512-1800). Request document S/N 052-003-01449-1.

Boettner, G. H., J. S. Elkinton, and C. J. Boettner. 2000. Effects of biological control introduction on three nontarget native species of saturniid moths. *Conservation Biology* 14: 1798-1806.

Bravo, A., Likitvivatanavong, S., Gill, S.S., Soberón, M., 2011. Bacillus thuringiensis: a story of a successful bioinsecticide. *Insect Biochem. Mol. Biol.* 41, 423–431.

C. Elvira, M. Goicoechea, L. Sampdro, S. Mato, R. Nogales, 1996. Bioconversion of solid paper-pulp mill sludge by earthworms, *Bioresour. Technol.* 75, 173–177.

C.A. Edwards, Earthworm Ecology, CRC Press, Boca Raton, FL, 2004.

Cornish PS, Stewart TEB. Certification – case studies with Australian market gardeners. In: Thompson, R. (ed.) Cultivating Communities. Proceedings of the 14th IFOAM Organic World Congress, 21–28 August 2002, Victoria, Canada. Canadian Organic Growers, Ottawa. 2002, 222.

Dabbert, S., Haring, A.M., Zanoli, R., 2004. Organic Farming Policies and Prospects. Zed Books, New York. 169 p.

E. Aranda, I. Barois, P. Arellano, S. Irisson, T. Salazar, J. Rodriguez, J.C. Patron, 1999. Vermicomposting in the tropics, in: P. Lavelle, L. Brussard, P. Hendrix (Eds.), Earthworm Management in Tropical Agroecosystems, CABI, UK, 253–287.

E. Benýtez, H. Sainz, R. Nogales, 2005. Hydrolytic enzyme activities of extracted humic substances during the vermicomposting of a lignocellulosic olive waste, Bioresour. Technol. 96 785–790.

Ecologically Based Pest Management (1996), a 260-page book published in 1996 by the National Academy of Science; available for $37.95. Phone 1-800-624-6242 to order. National Biological Control Institute.

Electronic Code of Federal Regulations, https://www.ecfr.gov/cgi-bin/text-idx?c=ecfr and sid= c4e0df8f46a4f4b6f56d80be31f95ed3 and rgn=div6 and view=text and node=7:3.1.1.9.32.4 and idno=7.

Elisa Morgera; Carmen Bullon Caro; Gracia Marin Duran, *Organic Agriculture and the Law*, http://www.fao.org/docrep/016/i2718e/i2718e.pdf, 2012.

Emerging Technologies for Integrated Pest Management: Concepts, Research and Implementation (2000), G.G. Kennedy and T.B. Sutton, eds. APS Press, St. Paul, MN. 526 p.

Federici, B.A., 2005. Insecticidal bacteria: an overwhelming success for invertebrate pathology. J. Invertebr. Pathol. 89, 30– 38.

Ferré, J., Van Rie, J., MacIntosh, S.C., 2008. Insecticidal genetically Modified crops and insect resistance management (IRM). In: Romeis, J., Shelton, A.M., Kennedy, G.G. (Eds.), Integration of Insect-Resistant Genetically Modified Crops within IPM Programs. Springer, Netherlands, pp. 41–85.

Finney, D. M., and N. G. Creamer. 2008. Weed management on organic farms. Center for Environmental Farming Systems organic production guide [Online]. North Carolina Cooperative Extension Service, Raleigh, NC. Available at: http://www.cefs.ncsu.edu/resources/organicproductionguide/weedmgmtjan808accessible.pdf (verified 11 March 2010).

Flint, K., Maria Louise and Dreistadt, Steve H. (1998). Natural Enemies Handbook: The Illustrated Guide to Biological Pest Control. University of California Press. ISBN9780520218017.

Get the Numbers on Organic, https://www.ccof.org/sites/default/files/Statistics per cent 202016.pdf, 2016.

Griffiths, G.J.K., Holland, J.M., Bailey, A., Thomas, M.B., 2008. Efficacy and economics of shelter habitats for conservation biological control. *Biological Control* 45, 200–209.

Grubinger, V. 1997. 10 steps toward organic weed control. American Vegetable Grower 46: 22–24.

Hammock B.D, (2001). Recombinant baculoviruses for insect control. *Pest Manag Sci.* 2001 Oct; 57(10):981-7.

Inceoglu AB, Kamita SG, Hinton AC, Huang Q, Severson TF, Kang K, Hammock B.D, (2001). Recombinant baculoviruses for insect control. *Pest Manag Sci.* 2001 Oct; 57(10):981-7.

J. Brodeur, J. Cory, J.D. Harwood, J.H. Hoffmann, B. Jacobsen, E. E. Lewis, P. Ode, D.O. TeBeest, S.B. Vinson, 2013. Biological Control Editorial Board. *Elsevier Journal* website. Copyright © 2013 Elsevier B.V.

James, C., 2010. Global Status of Commercialized Biotech/GM Crops: 2010. International Service for the Acquisition of Agri-Biotech Applications. <http://www.isaaa.org/resources/publications/briefs/42/default.asp> (accessed 29.02.12).

Janssen, M. and Hamm, U., 2012. Product labelling in the market for organic food: Consumer preferences and willingness-to-pay for different organic certification logos. Food quality and preference, 25(1), pp.9-22.

Johnson M., Wratten S.D., Landis D.A. and Gurr, G.M., 2008. Recent advances in conservation biological control of arthropods by arthropods. *Biological Control* 45, 172–175.

Karen A. O'Hanlon, Kamilla Knorr, Lise Nistrup J rgensen, Mogens Nicolaisen, Birte Boelt, 2012. Exploring the potential of symbiotic fungal endophytes in cereal disease suppression. *Biological Control* 63 (2012) 69–78 Review.

Kginter, *How to Get a Certified Organic Product Label (in the USA),* http://blog.quicklabel.com/2010/10/how-to-get-a-certified-organic-product-label-in-the-usa/, October 26, 2010.

Kirchmann H, Ryan MH. Nutrients in organic farming – are there advantages from the exclusive use of organic manures and untreated minerals. In: Fischer, T., Turner, N., Angus, J., McIntyre, L., Robertson, M., Borrell, A. and Lloyd, D. (eds). New directions for a diverse planet. Proceedings of the 4th International Crop Science Congress. 26 Sept.–1 Oct. 2004, Brisbane, Australia. The Regional Institute, Gosford. pp. CD ROM, 2004.

Kogan M (1998) Integrated pest management: historical perspectives and contemporary developments. *Annu Rev Entomol* 43:243–270

Kost, T. A., Condreay, J. P., and Jarvis, D. L. 2005. Baculovirus as versatile vectors for protein expression in insect and mammalian cells. *Nature Biotech.* 23:567-575. L. A.

L. A. Lacey, R. Frutos, H. K. Kaya, and P. Vail, 2001.Insect Pathogens as Biological Control Agents. Do They Have a Future? *Biological Control* 21, 230 –248 (2001).

L.P.S. Chan, D.A. Griffiths, 1998. The vermicomposting of pre-treated pig manure, *Biol.Wastes* 24, 57–69.

Labeling Packaged Products Under the National Organic Standards, https://www.ams.usda.gov/sites/default/files/media/NOP per cent 20Labeling per cent 20 Packaged per cent 20Products.pdf, July 27, 2009.

Lacey, L.A., R. Frutos, H.K. Kaya and P. Vails. (2001). Insect pathogens as biological control agents: Do they have future? *Biol. Contr.,* 21(3): 230- 248.

Lacey, R. Frutos, H. K. Kaya, and P. Vail, 2001.Insect Pathogens as Biological Control Agents. Do They Have a Future? *Biological Control* 21, 230 –248 (2001).

Landis, D.A., Wratten, S.D., Gurr, G.M., 2000. Habitat management to conserve natural enemies of arthropod pests in agriculture. *Annual Review of Entomology* 45, 175–201.

Lee, J.M., Kojin, Y., Tatsuke, T., Mon, H., Miyagawa, Y. and Kusakabe, T.(2013) Co expression of *Escherichia coli* RNase III in silkworm cells improves the efficiency of RNA interference induced by long hairpin dsRNAs. *Insect Science*, 20, 69–77.

Liebman, M., and E. R. Gallandt. 1997. Many little hammers: ecological approaches for management of crop–weed interactions. p. 291–343. *In* L. E. Jackson (ed.) Ecology in agriculture. Academic Press, San Diego, CA.

M.S. Hoddle, 2003. CLASSICAL BIOLOGICAL CONTROL OF ARTHROPODSIN THE 21ST CENTURY. 1st International Symposium on Biological Control of Arthropods. Department of Entomology, University of California, Riverside, California, U.S.A.

M.S. Maboeta, L. Van Rensburg, 2003. Vermicomposting of industrially produced wood chips and sewage sludge using *E. foetida, Ecotoxicol. Eniron. Saf.* 56, 256–270.

Mao, Y.B., Cai, W.J., Wang, J.W., Hong, G.J., Tao, X.Y., Wang, L.J., Huang, Y.P. and Chen, X.Y. (2007) Silencing a cotton bollworm P450 monooxygenase gene by plant-mediated RNAi impairs larval tolerance of gossypol. *Nature Biotechnology,* 25, 1307–1313.

McCaffery, A.R., 1998. Resistance to insecticides in heliothine Lepidoptera: a global view. Philosophical Transactions of the Royal Society of London Series B-Biological Sciences 353, 1735–1750.

Microbial Control of Weeds (1991), D.O. TeBeest, ed. Chapman and Hall. New York. 284 p.

Mohler, C. Undated. Ecology of weed management in organic systems [Online slide presentation]. Available at: http://www.css.cornell.edu/WeedEco/Mohler *Ecology of Weed manag Presentation*.pdf (verified 11 March 2010).

Morin, L., Reid, A.M., Sims-Chilton, N.M., Buckley, Y.M., Dhileepan, K., Hastwell, G.T., Nordblom, T.L., Raghu, S., 2009. Review of approaches to evaluate the effectiveness of weed biological control agents. *Biological Control* 51, 1–15.

Moscardi, F. 2007. A Nucleopolyhedrovirus for control of the velvet bean caterpillar in Brazilian Soybeans. In Biological Control: A Global Perspective, eds. C. Vincent, M. S. Goethel, and G. Lazarovits, pp. 344-352. Oxford shire, UK, and Cambridge, USA: CAB International.

Mráèek Z, Jiskra K, Kahounova L. Efficiency of steinernematid nematodes (Nematoda: Steinernematidae) in controlling larvae of the black vine weevil, *Otiorrhynchus sulcatus* (Coleoptera: Curculionidae), in laboratory and field experiments. *European Journal of Entomology.* 1993; 90:71–76.

Nguyen, N. C. Borgemeister, H. Poehling and G. Zimmermann (2007). Laboratory investigations on the potential of entomopathogenic fungi for biocontrol of Helicoverpa armigera (Lepidoptera: Noctuidae) larvae and pupae. *Biocontrol Sci. Technol.*, 17(12): 853-864.

Organic Certification and Accreditation, https://www.ams.usda.gov/services/organic-certification.

Organic Labeling Requirements, http://www.nsf.org/consumer-resources/green-living/organic-certification/organic-labeling-requirements.

Organic Labeling, https://www.ams.usda.gov/rules-regulations/organic/labeling

P. Sangwan, C.P. Kaushik, V.K. Garg, (2008). Feasibility of utilization of horse dung spiked filter cake in vermicomposters using exotic earthworm *Eisenia foetida, Bioresour. Technol.* 99 2442–2448.

Padel, S. and Lampkin, N. eds., 1994. The economics of organic farming: an international perspective (pp. 201-222). Wallingford: Cab International.

Parra JRP, Botelho PSM, Corrêa-Ferreira BS, Bento JMS (2002) Controle Biológico no Brazil: parasitóides e predadores. Editora Manole, S o Paulo. 609p

Pattison, R.R., Mack, R.N., 2007. Potential distribution of the invasive tree Triadica sebifera (Euphorbiaceae) in the United States: evaluating climex predictions with field trials. *Global Change Biology* 14, 813–826.

Pfiffner, L., Wyss, E., 2004. Use of sown wildflower strips to enhance natural enemies of agricultural pests. In: Gurr, G.M., Wratten, S.D., Altieri, M.A. (Eds.), Ecological Engineering for Pest Management: Advances in Habitat Manipulation for Arthropods. CSIRO Publishing, Collingwood, pp. 165–186

Pickett, C.H., Simmons, G.S., Lozano, E., Goolsby, J.A., 2004. Augmentative biological control of whiteflies using transplants. *Biocontrol* 49, 665–688.

Price, D.R. and Gatehouse, J.A. (2008) RNAi-mediated crop protection against insects. *Trends in Biotechnology*, 26, 393– 400.

Purwar, J. P. and G.C. Sachan. (2005). Biotoxicity of *Beauveria bassiana* and *Metarhizium anisopliae* against *Spodoptera litura* and *Spilarctia oblique. Ann. Pl. Protec. Sci.*, 13(2): 360-364.

R. Gupta, V.K. Garg, 2008. Stabilization of primary sludge during vermicomposting, *J. Hazard. Mater.* 153, 1023–1030.

Richard, A., Lankau1, R.A., Rogers, W.E., Siemann, E., 2004. Constraints on the utilization of the invasive Chinese tallow tree *Sapium sebiferum* by generalist native herbivores in coastal prairies. Ecological Entomology 29, 66–75.

S. H. Dreistadt, 2007. Biological control and Natural enemies. UC Statewide IPM Program, Davis. Produced by UC Statewide IPM Program, University of California, Davis

Sauka, D.H., Benintende, G.B., 2008. Bacillus thuringiensis: generalidades. Unacercamiento a su empleo en el biocontrol de insectos lepidópteros que son plagas agrícolas. *Rev. Argent. Microbiol.* 40, 124–140.

Shapiro-Ilan, D. I., T. E. Cottrell, R. F. Mizell, D. L. Horton, B. Behle, and C. Dunlap. 2010. Efficacy of Steinernema carpocapsae for control of the lesser peachtree borer, Synanthedon pictipes: Improved aboveground suppression with a novel gel application. Biological Control 54, 23–28.

Shelton, A.M., Zhao, J.Z., Roush, R.T., 2002. Economic, ecological, food safety, and social consequences of the deployment of Bt transgenic plants. *Annu. Rev. Entomol.* 47, 845–881.

Shokoofeh Kamali, Javad Karimi, Mojtabi Hosseini, Requel Compos -Herrera and Larry W. Duncan (2013). Biocontrol potential of the entomopathogenic nematodes Heterorhabditis acteriophora and Steinernema carpocapsae on cucurbit fly, Dacus ciliatus (Diptera: Tephritidae). Biocontrol Science and Technology Volume 23, Issue 11, 2013. Pages 1307-1323.

Stuart, R. J., M. E. Barbercheck, P. S. Grewal, R.A.J. Taylor, and C. W. Hoy. 2006. Population biology of entomopathogenic nematodes: Concepts, issues, and models. *Biological Control* 38: 80-102.

Sullivan, P. 2003. Principles of sustainable weed management for croplands [Online]. ATTRA Publication #PO39. National Sustainable Agriculture Information Service. Available at: https://attra.ncat.org/attra-pub-summaries/?pub=109 (verified 4 June 2019).

Trewavas A. A critical assessment of organic farming and-food assertions with particular respect to the UK and the potential environmental benefits of no-till agriculture. *Crop Protection.* 2004; 23(9):757-781.

Trewavas A. Urban myths of organic farming. *Nature.* 2001; 410:409-410.

U. Reeh, 1992. Influence of population densities on growth and reproduction of the earthworm Eisenia andrei on pig manure, Soil Biol. Biochem. 24, 1327–1331. [14] S. Gajalakshmi, E.V. Ramasamy, S.A. Abbasi, 2002. Vermicomposting of different forms of water hyacinth by the earthworm Eudrilus eugeniae, Kinberg, *Bioresour. Technol.* 82, 165–169.

Use of Plant Pathogens in Weed Control (1999), E.N. Rosskopf, R. Charudattan and J.B. Kadir. In: Handbook of Biological Control, Academic Press, San Diego, CA p. 891-918.

V.K. Garg, P. Kaushik, 2005. Vermistabilization of textile mill sludge spiked with poultry droppings by an epigeic earthworm *Eisenia foetida, Bioresour. Technol.* 96, 1189–1193.

V.K. Garg, P. Kaushik, N. Dilbaghi, 2005. Vermiconversion of wastewater sludge from textile mill spiked with anaerobically digested biogas plant slurry employing *Eisenia foetida, Ecotoxicol. Environ. Saf.* 65 (3), 412–419.

Van Driesche, R.G., Lyon, S., Sanderson, J.P., Bennett, K.C., Stanek, E.J., Zhang, R.T.,2008. Greenhouse trials of Aphidius Colemani (Hymenoptera: Braconidae) banker plants for control of aphids (Hemiptera: Aphididae) in greenhouse spring floral crops. *Florida Entomologist* 91, 583–591.

Van Lenteren JC (2000) Critérios de sele o de inimigos naturais a serem usados em programas de controle biológico. In: Bueno VHP (ed) Controle biológico de pragas: produ o massal econtrole de qualidade. Editora UFLA, Lavras, 196p, pp. 1–19.

Van Lenteren, J. C. 1983. Biological pest control: passing fashion or here to stay? Organorama (Netherlands) 20: 1-9.

Vincent Vachon, Raynald Laprade, Jean-Louis Schwartz, 2012. Current models of the mode of action of Bacillus thuringiensis insecticidal Crystal proteins: A critical review. *Journal of Invertebrate Pathology* 111 (2012) 1–12.

Wang, Y.L., Yang, M.L., Jiang, F., Zhang, J.Z. and Kang, L. (2013a) MicroRNA-dependent development revealed by RNA interference-mediated gene silencing of LmDicer1 in the migratory locust. *Insect Science*, 20, 53–60.

What is Organic? https://www.ccof.org/organic.

Yan, X., Han, R.C., Moens, M., Chen, S.L. and De Clercq, P. 2013. Field evaluation of entomopathogenic nematodes for biological control of striped flea beetle, *Phyllotreta striolata* (Coleoptera: Chrysomelidae) *Biocontrol* 58: 247-256.

Zhang, X.Y., Lu, K., Zhou, J.L. and Zhou, Q. (2013b) Molecular characterization and gene functional analysis of Dicer2 gene from *Nilaparvata lugens* (Hemiptera: Geometroidea). *Insect Science*, 20, 61–68.

Zhuang, J.J. and Hunter, C.P. (2012) RNA interference in *Caenorhabditis elegans*: uptake, mechanism, and regulation. *Parasitology*, 139, 560– 573.

Index

C

D

E

F

9 789359 191874

Printed at Repro India Ltd.